Software Performance Risk Management

Engineering Survival in a World Built on Fragile Software

First Edition

James L. Pulley III

`https://www.journeymanpublishing.com`

First Edition, 2026.

Library Edition	978-1-964222-48-6
Paperback:	978-1-964222-56-1
eBook:	978-1-964222-17-2
Hardback:	978-1-964222-16-5

Library editions are manufactured to institutional durability standards.

Journeyman Publishing is a specialty publisher for Cybersecurity, IT Professional Services, Software Performance Engineering, Technical Selling, & Software Quality Assurance books. If you have an interest in becoming a published writer, please contact us at newauthor@journeymanpublishing.com

Library of Congress Cataloging-in-Publication Data
Names: Pulley, James Leonard, III, author.
Title: Software performance risk management : governing survivability, structural exposure, and business risk in modern software systems / James Leonard Pulley III.
Description: Pauline, South Carolina : Journeyman Publishing LLC, 2026 | Includes bibliographical references and index.
Identifiers: LCCN 2026937354 - ISBN 978-1-964222-48-6 (library binding) 978-1-964222-16-5 (casebound) 978-1-964222-56-1 (perfect binding) 978-1-964222-17-2 (ebook)
Subjects: LCSH: Software engineering—Risk management. Information technology—Governance. Business enterprises—Risk management. Computer systems—Failure analysis. Organizational resilience.
Classification: LCC QA76.76.R5 P85 2026 DDC 005.1—dc23
LC record available at: https://lccn.loc.gov/2026937354

Journeyman Publishing

111 Foster Mill Circle, Pauline South Carolina 29374
https://www.journeymanpublishing.com

Foreword by Dennis Hayes

A few years ago, I decided I wanted to attend the Spartanburg County Republican Party organizing convention. I wasn't going for a headline or a speech. I simply wanted to observe the process up close and personal — to see how it worked from the inside.

Because I have been blind for several years, attending something like that requires a bit of planning. I posted on Facebook to see whether someone already going might be willing to let me ride with them and help me navigate the building once we arrived. I didn't want special treatment — just practical assistance so I could participate like anyone else.

James Pulley's wife, Rachel, saw my post. She asked James to give me a call. Within a short time, we had arranged to ride together, and that small act of generosity marked the beginning of one of the best friendships I have known.

From that day forward, James and Rachel became not just colleagues, but trusted friends. They stood as witnesses at my wedding to Cathy McBride at the end of July 2025 — a moment that meant more to Cathy and me than words can fully express. Friendships forged through shared purpose, respect, and intellectual curiosity tend to endure, and ours has been exactly that.

Since that first convention ride, James and I have had many long conversations about politics, philosophy, and computer science. When I founded the American Technology Venture Lab in Spartanburg, South Carolina — an objective-based mentoring and educational program for early-stage science and technology ventures — James joined the board and has consistently been thoughtful, strategic, and generous with his time. He has also been deeply supportive of Spartanburg Supports Law Enforcement and First Responders, Inc., doing business as SPARTANSFIRST. In both organizations, his perspective has been measured, principled, and practical.

Many of our conversations drifted into the domain where James is best known among his peers: software performance.

Occasionally we would talk about what happens when someone appears for the first time on Fox News, or when a founder is invited onto Shark Tank, and suddenly their systems experience extreme traffic loads. Websites crash. APIs stall. Payment systems fail. What looks like success from the outside can become technical collapse on the inside.

James would patiently explain aspects of performance engineering that I understood only in outline. Fortunately, my background in computer science at Georgia Tech and my decades in the technology industry gave me enough foundation to follow the architecture of what he was describing.

During my years at Hayes Microcomputer Products — when we developed the Hayes Smartmodem and the AT command set — I learned a great deal about rigorous testing and quality control. Every time we introduced a faster modem or a new hardware platform, we had to ensure that the AT command structure remained consistent and reliable. That discipline gave me a deep appreciation for standards, repeatability, and structured validation.

But performance engineering at enterprise scale — especially in the modern distributed, dependency-heavy world — is a different domain. Over the past several years, through my discussions with James, I often felt as though I had enrolled in a graduate-level seminar in performance engineering from a strategic management perspective. I may not have become an implementation expert, but I came to understand the deeper implications: performance is not merely about speed — it is about survivability, governance, and enterprise risk.

When James told me he was writing this book, I was genuinely pleased. For decades, many of his peers have referred to him — with respect and affection — as the "godfather of performance testing." That nickname fits. Not only does he think outside the traditional box, but he also understands how to teach what he knows — not just the technical mechanisms, but the strategic business implications.

This book demonstrates that rare combination.

One of the most powerful insights in these pages is the distinction between visibility and governance. In Chapter 1, James dismantles what he calls "The Illusion of Control" — the mistaken belief that dashboards and instrumentation equal safety. Visibility observes. Governance decides. That distinction alone reframes how executives should think about digital infrastructure.

In Chapter 2, his discussion of performance testing as "fictional certainty" is not a dismissal of testing — far from it. It is an elevation of it. Performance testing creates conditional certainty within known boundaries. But as James makes clear, DNS failures, certificate trust collapse, routing instability, and third-party dependency concentration exist beyond the boundary of what traditional performance testing can safely simulate. That boundary is where survivability risk begins.

The introduction of the Risk Equation in Chapter 10 — combining initiation uncertainty, blast radius, and recovery feasibility — shifts the conversation from technical metrics to structural exposure. It forces leaders to ask not just "Will it run?" but "How much breaks when it fails?" and "Can we recover?" That is governance language, not developer language.

His chapters on Hidden Foundations and Trust as Infrastructure (Chapters 11 and 15) are especially important in an era of SaaS concentration, CDN dependency, identity providers, and protocol-layer fragility. These are not hypothetical concerns. They are architectural realities that most enterprises depend upon without mapping or governing them explicitly.

What James has done in this book is give a name — Software Performance Risk Management — to something that has existed implicitly for years but has never been formally defined. He recognizes that performance, survivability, and enterprise risk are inseparable in a world built on fragile software systems.   By naming the discipline, he gives organizations the language and authority structure necessary to govern it.

This is not a book about tuning code. It is not a manual for monitoring tools. It is a call for executive ownership of survivability.

For major enterprises in America and throughout the world — where digital continuity determines economic continuity — this framework is not optional. It is foundational.

I have known James Pulley as a friend, a board member, a teacher, and a thinker. I have watched him translate deeply technical realities into business strategy and governance principles. This book reflects that rare ability.

I am proud to call him my friend. And I am grateful that he has chosen to share his insight with the broader industry.

Dennis C. Hayes

Field Observations

Software systems do not fail randomly. They fail along paths that were never measured.

They fail through assumptions that were never questioned. They fail through dependencies that were never mapped. They fail through trust that was never verified.

The industry has spent decades perfecting how to make software fast. Almost no one has asked how to make it survive.

We built tools to observe applications. We did not build tools to observe fragility.

We optimized performance. We ignored exposure.

We tuned response times. We ignored blast radius.

We instrumented code. We never instrumented risk.

Software Performance Risk Management is not a new metric. It is a new responsibility.

It exists because complexity crossed a threshold that intuition can no longer manage.

It exists because survivability is more important than speed.

It exists because confidence without risk intelligence is just ignorance with better charts.

This book is not about improving performance.

It is about preventing collapse.

Software Performance Risk Management (SPRM) — Discipline Definition

Software Performance Risk Management (SPRM) is a formal discipline concerned with the identification, measurement, governance, and mitigation of performance-related risks that threaten the reliability, economic viability, and trustworthiness of software-dependent systems.

SPRM treats performance not as a late-stage optimization activity, nor as a by-product of testing or observability, but as a persistent risk surface that exists across the full lifecycle of software systems—design, construction, integration, deployment, operation, and evolution.

The discipline is grounded in the following principles:

Performance is a business risk, not a technical preference. Degradation in latency, throughput, availability, or scalability manifests as financial loss, reputational damage, regulatory exposure, and operational fragility.

Risk precedes measurement; measurement precedes optimization. SPRM prioritizes understanding where performance failure matters before deciding how to measure it and long before attempting to optimize it.

Absence of failure evidence is not evidence of absence. SPRM assumes that unmeasured or unobserved performance characteristics represent unknown risk, not safety.

Production reality outweighs synthetic certainty. The discipline values empirical evidence from real execution paths, real dependencies, and real user behavior over idealized or isolated test conditions.

Performance risk propagates across boundaries. Third-party services, networks, identity systems, time sources, control planes, and external platforms are treated as first-class contributors to performance exposure.

Governance is as critical as instrumentation. SPRM establishes ownership, accountability, thresholds, and escalation paths for performance risk, independent of specific tools or vendors.

SPRM does not replace performance testing, observability, reliability engineering, or capacity planning. Instead, it unifies and governs them under a risk-oriented framework, ensuring that performance effort aligns with material business exposure rather than convenience, habit, or tooling availability.

What SPRM Is Not

- Not a testing methodology
- Not an APM strategy
- Not a monitoring stack
- Not a benchmarking exercise
- Not a vendor category

SPRM is a discipline of intent, prioritization, and governance, capable of being implemented with many toolchains – or with few.

Software Performance Risk Management exists because accountability for survivability did not.

SPRM asserts that survivability must be governed with the same executive authority as financial risk, because failure without accountability is not engineering – it is negligence.

Preface

This book exists because modern software systems reached a point where performance could no longer be separated from survival.

For decades, performance engineering focused on speed, efficiency, and scalability. Those efforts succeeded. Systems became faster, more distributed, and more complex than ever before.

They also became fragile in ways no existing tool was designed to measure.

Outages grew larger. Recovery grew slower. Dependencies grew deeper. Trust grew thinner.

And yet our language for performance never changed.

Software Performance Risk Management exists to correct that imbalance. It reframes performance as a business survival risk and introduces the models, language, and discipline required to manage it.

Software Performance Risk Management exists because no existing discipline is accountable for survivability before failure.

Performance engineering optimizes systems that already exist. Observability explains systems after they execute. Reliability engineering improves systems once they are allowed into production. Incident response manages damage after it occurs. None of these disciplines are responsible for governing whether architectural decisions, dependency concentration, or trust assumptions are survivable before they are deployed at scale. That gap is not technical. It is a failure of accountability. SPRM exists to close it.

This book is written for:

- Engineers who already know how to make systems fast
- Architects who must now make systems survivable
- Executives who must govern digital continuity

It does not replace performance engineering. It completes it.

Some case examples in this work are representative narratives synthesized from publicly documented incidents and industry patterns, presented to illustrate structural risk dynamics rather than to provide exhaustive historical accounts.

How to Read This Book

This work is intended for professionals responsible for the design, operation, and governance of software-dependent systems where performance failure represents material business risk.

This book does not introduce tools, dashboards, agents, or metrics. If you are looking for a new observability platform, a performance testing methodology, or a reporting framework, this book will disappoint you. Software Performance Risk Management is not a tooling discipline. It is a governance discipline. Its purpose is not to improve how systems are observed, but to determine which risks are acceptable before systems are allowed to operate at scale.

Part I dismantles the belief that the current performance and observability model is sufficient. It shows why even perfect execution visibility cannot see survivability risk.

Part II defines Software Performance Risk Management as a formal discipline, introducing its language, models, and foundations.

Part III turns SPRM into an operational program that can be executed inside real organizations.

This is not a book to skim. It is a book to work with.

If you are uncomfortable with questions of authority, accountability, and ownership, this book will challenge you more than it instructs you.

Question assumptions. Mark where your architecture is exposed.

That is how disciplines are born.

Contents

III Building the Safe Harbor 127

16 Stewardship 131

17 From Observed to Measured: Escaping the Dashboard Ceiling 137

18 From Measured to Risk Aware 145

IV Failure Under Load: Case Studies in Ungoverned Survivability   177

22 Crisis Load: When Systems Are Forced Outside Their Design Envelope   181

23 The Cost of Not Knowing   187

Part I

Why Everything We Use to Measure Performance Is Structurally Incomplete

The Illusion of Control

Visibility does not create safety. It creates confidence. And confidence is often the most dangerous state a system can be in.

Modern engineering organizations are saturated with dashboards. They measure latency, throughput, error rates, and saturation. Each one implies the same belief: *If we can see it, we can manage it.* That belief is false.

Foundational Definition: Control vs. Governance

Visibility is the ability to observe system behavior after execution begins.

Control is the ability to influence system behavior during execution.

Governance is the authority to determine which behaviors, dependencies, and risks are allowed to exist before execution occurs.

Modern software organizations have mistaken visibility for governance. This error produces confidence without authority—and confidence without authority is exposure.

Highly visible systems fail just as catastrophically as invisible ones, sometimes more so, because visibility creates the illusion that risk is being managed when it is merely being observed.

When organizations say performance is under control, they usually mean the application is instrumented, tests passed, monitors are green, and dashboards are quiet. None of these describes survivability. They describe internal order.

Dashboards measure activity within authorized systems.

They do not measure whether those systems should be authorized to exist in their current form. No dashboard can show dependency concentration, trust fragility, or blast radius. Those properties are architectural, not operational. The moment an organization expects dashboards to provide governance, it has already delegated authority to systems that cannot exercise it.

They say nothing about DNS resolution, routing stability, certificate validation, CDN availability, email trust, or dependency collapse. These are ecosystem problems, not application problems.

These failures persist not because they are unknown, but because they fall between ownership boundaries.

No performance team governs DNS survivability. No application team owns routing convergence. No observability platform is accountable for certificate trust collapse. These risks exist outside traditional domains of responsibility, which means they are governed by assumption rather than authority.

Boundary Clarification

Not all failures are governance failures. Many outages are the result of defects, misconfigurations, or operational error, and engineering disciplines exist precisely to reduce their frequency and duration.

A failure becomes a governance failure only when its blast radius, recovery feasibility, or dependency concentration was knowable in advance — and no authority was accountable for deciding whether that risk was acceptable.

SPRM does not claim that governance can prevent all failure. It claims that unowned survivability risk guarantees catastrophic failure over time.

Performance was defined too narrowly. Speed does not describe fragility. Speed does not describe blast radius. Speed does not describe survivability.

The illusion of control does not come from ignorance. It comes from misapplied authority.

When organizations treat visibility as governance, they outsource survivability to chance. When they treat performance as speed alone, they ignore exposure. Software Performance Risk Management begins with a single correction: visibility explains systems, but authority governs them. Until survivability has an owner, collapse remains a statistical certainty.

2

Performance Testing: Fictional Certainty

Performance testing was born from necessity.

As software systems moved from single-user applications to shared services, then to web platforms, and finally to global digital infrastructure, failure changed shape. It was no longer acceptable for software to merely function. It had to function under pressure.

Performance testing exists to answer one fundamental question:

Does this system behave correctly when real-world load is applied?

It was never about speed alone. It was about collapse prevention.

Performance testing does not produce false certainty. It produces conditional certainty.

The certainty it provides is real, engineered, and essential—but it is valid only within the boundaries of the world assumed by the test. When those boundaries are crossed, certainty ends. Software Performance Risk Management begins where those boundaries become visible.

Early performance testing focused on:

- Concurrency limits
- Resource saturation
- Deadlocks
- Memory exhaustion
- Connection starvation

These were catastrophic failure modes in early distributed systems. Performance testing prevented them. That alone justifies its existence as a core engineering discipline.

Performance Testing as a Pillar of Quality Assurance

Performance testing operates entirely within the boundary of the system under test.

It assumes the existence and cooperation of infrastructure, networks, identity systems, time synchronization, and external providers. These assumptions are not oversights—they are prerequisites. A test cannot execute in a world that does not function. As a result, performance testing cannot govern failures that occur outside those assumptions, no matter how well the test is designed or executed.

Performance testing sits inside QA because QA exists to protect the business from unknown failure.

Functional testing asks: *Does it work?*

Security testing asks: *Can it be compromised?*

Performance testing asks: *Will it collapse under legitimate use?*

This places performance testing in a unique position: It is the only QA discipline that protects the organization from *scale-induced failure*.

Without performance testing:

- Systems pass unit tests and still collapse
- Features ship that cannot survive success

- Growth becomes existential risk

Performance testing is how success is made safe.

What Performance Testing Does Exceptionally Well

Performance testing is unmatched at:

- Identifying bottlenecks
- Revealing scaling ceilings
- Exposing inefficient algorithms
- Stressing infrastructure capacity
- Validating architectural assumptions

It is the only discipline that forces engineers to confront reality.

It turns theoretical architecture into physical behavior.

This is engineering maturity.

Hero Story: From Ten Thousand to a Nation

In September of 2021, El Salvador attempted something no country had ever done before. It launched a national digital currency wallet based on Bitcoin. The system was called *Chivo*.

The first launch failed.

Publicly. Painfully. At national scale.

President Nayib Bukele addressed the nation and acknowledged something most leaders rarely say out loud:

"Software is hard."

He compared the failure to *healthcare.gov* in the United States and the UK National Health Service rollout. Large ambition, large complexity, and large consequences.

El Salvador did not abandon the idea. They abandoned the implementation.

Later that fall, a new vendor was selected and a new system was designed from the ground up. There was no production telemetry. No usable operational history. No performance baselines. Nothing that could safely be transferred from the failed platform.

From a performance engineering perspective, the system was a blank slate.

That is not a disadvantage. That is the purest form of performance testing.

When no real-world data exists, reality must be manufactured. Performance testing is how reality is created before it exists.

The new system, in its earliest form, could not scale past ten thousand concurrent users.

At that point:

- Error rates surged
- Timeouts dominated request flows
- Response times became unstable
- User trust evaporated

This was not a tuning problem. It was a survivability problem.

Without intervention, the system was structurally incapable of supporting a national rollout.

A skilled performance testing team worked alongside architects and developers. Their role was not to criticize, but to construct a controlled, repeatable version of national reality.

They did three things:

1. Built synthetic load models that mirrored national adoption behavior

2. Executed those models relentlessly and consistently

3. Converted absence of production data into engineered truth

The performance test harness became a national simulator.

Every test run produced:

- Timing distributions
- Throughput ceilings
- Error collapse points
- Resource exhaustion signatures

Across:

- CPU
- Memory
- Disk
- Network

This was not speculation. It was physics.

What followed was not a single fix. It was a cascade of insight.

The performance testing team worked directly with application developers and system architects. Together they identified:

- Hidden contention points
- Serialization bottlenecks
- Invalid scaling assumptions
- Structural concurrency limits

Each discovery became engineering action:

- Caching strategies were redesigned
- Database access patterns were refactored
- Thread pools were rebalanced
- Dependency paths were shortened
- Resource contention was eliminated

And then the system was tested again.

And again.

And again.

Performance testing became a feedback engine, not a validation step.

In four weeks, the system moved:

From ten thousand concurrent users To over one million concurrent users.

That is not optimization. That is transformation.

Late nights followed. Weekends disappeared. Everyone understood what failure meant at national scale.

This was not a website. This was monetary infrastructure.

When the second launch arrived, the system held.

Not because of hope. Not because of luck. Because capacity had been engineered into existence.

This success was possible because the system was still architecturally pliable. Dependencies could be redesigned. Assumptions could be challenged. Trust relationships were not yet fixed. Performance testing can only protect systems while these conditions remain true. Once dependencies harden and control shifts outside the organization, survivability becomes a governance problem, not an engineering one.

This is performance testing at its highest form.

Not a QA checkbox. Not a tuning exercise. Not a report.

It is the difference between:

"We think this will work."

And:

"We have proven it will survive."

But its boundary must be understood.

Performance testing saved this system because:

- The architecture was still changeable
- Dependencies were still malleable
- Survivability could still be designed

Performance testing operates inside architecture. It does not govern the infrastructure that surrounds it.

It can ensure the ship floats. It does not control the ocean.

That is not its weakness.

It is its purpose.

Performance testing is the discipline that turns theoretical architecture into physical truth.

No other practice forces systems to reveal their breaking points before users discover them. This is precisely why performance engineers should not be held accountable for failures that occur beyond those breaking points. Their discipline exposes what can be known. Software Performance Risk Management exists to govern what cannot be tested safely, economically, or at all.

The Difficulty of Minting Performance Engineers

Performance engineers are not created through certification.

They are created through:

- Failure exposure
- Deep system intuition
- Cross-discipline understanding

They must understand:

- Code
- Infrastructure
- Networking
- Databases
- Operating systems

This makes performance engineering one of the hardest disciplines to staff.

Organizations often:

- Promote functional testers into performance roles
- Assign developers part-time
- Outsource critical testing phases

This creates fragility not in tooling, but in expertise.

SPRM does not replace performance engineers. It elevates their work by giving it business context.

Performance Testing as Business Risk Mitigation

Performance testing reduces business risk by:

- Preventing revenue loss due to overload
- Protecting brand reputation
- Avoiding emergency scaling costs
- Preventing regulatory exposure

It transforms engineering correctness into business stability.

But the risk it mitigates is narrow:

Risk caused by the system misbehaving when it is asked to operate.

The Structural Boundary

Performance testing assumes:

- DNS resolves
- TLS validates
- Routing functions
- Providers cooperate
- Time behaves

These assumptions define the jurisdictional boundary of performance testing. Any failure mode that violates them exists outside the authority of the discipline. Holding performance testing accountable for survivability beyond this boundary is not rigorous – it is negligent. Software Performance Risk Management exists to make these boundaries explicit, govern what lies beyond them, and prevent organizations from mistaking tested behavior for guaranteed survival.

These conditions must be true for the test to run.

Which means they cannot be tested.

Performance testing requires a stable world. Survivability is defined by an unstable one.

This is not a flaw. It is a boundary.

Why Performance Testing Creates Fictional Certainty

Passing performance tests increases confidence.

But that confidence is conditional: *The system works if the world cooperates.*

Most catastrophic outages occur when the world does not cooperate.

That is why performance testing creates certainty that is real, useful, and incomplete.

Performance testing does not fail when systems collapse outside their tested conditions.

Organizations fail when they treat conditional certainty as absolute authority. Performance testing makes success possible. Software Performance Risk Management makes success survivable. Confusing the two is how well-tested systems still fail catastrophically.

3

APM: Deep Vision, Shallow Horizon

Before APM, production systems were black boxes.

Operations teams monitored hosts:

- CPU
- Memory
- Disk
- Network

These metrics described infrastructure health, not application behavior.

When systems were simple and monolithic, this was enough. When systems became distributed, it was not.

Failures began occurring inside:

- Application code
- Databases
- Message queues
- External APIs

Infrastructure metrics remained green while applications collapsed.

APM was created to close that gap.

It was the first discipline to say: *We must observe software as it executes, not just the machines that host it.*

APM as a Response to Architectural Complexity

As architectures evolved:

- Monoliths became services
- Services became graphs
- Dependencies multiplied

Failures became:

- Partial
- Cascading
- Non-deterministic

APM introduced:

- Transaction tracing
- Cross-service visibility
- Latency attribution

It transformed "something is slow" into "this function is slow because that dependency is slow."

This was revolutionary.

The limitation of APM is not depth, precision, or accuracy. It is horizon.

APM observes systems only after execution begins. Its visibility extends precisely as far as the system is able to run. When execution never starts — because trust fails, routing fractures, or dependencies collapse — APM has nothing to observe. This is not a flaw. It is the boundary of execution-based visibility.

What APM Does Exceptionally Well

APM excels at:

- Root cause analysis
- Latency breakdown
- Capacity diagnosis
- Code-level accountability

It answers:

- Where time is spent
- Where errors originate
- Which services degrade first

APM creates engineering precision.

APM is a tactical control system.

It shortens outages, accelerates diagnosis, and restores service. It governs execution behavior while systems are running. What it does not do, and was never designed to do, is determine whether the risks embedded in architecture, dependency selection, or trust relationships are acceptable before execution occurs.

Hero Story: When MTTR Became Business Risk

For William Hill[1], one of the world's largest sports betting companies, mean time to resolve (MTTR) was not an operational metric. It was a business survival metric.

Every minute of degraded performance meant:

- Lost transactions
- Lost customer trust
- Lost revenue

---

[1]William Hill case study published by New Relic, describing observability consolidation and an 80% reduction in MTTR: `https://newrelic.com/blog/apm/customers-apm-monitoring`

Before adopting a unified APM approach, their observability stack was fragmented. Different teams used different tools. Data was scattered. Correlation was slow. Diagnosis was slower. When failures occurred, engineers spent precious time stitching together partial truths from disconnected systems.

Risk lived in the gaps.

By consolidating application, infrastructure, and service telemetry into a single APM platform using New Relic, William Hill fundamentally changed how incidents were understood and resolved. Performance data, error traces, and system behavior could now be seen in one place, in real time.

The result was dramatic.

MTTR dropped by more than 80%, and reliability approached 100%.

This outcome demonstrates APM at its highest form: accelerating recovery after failure.

It does not demonstrate risk prevention. Faster recovery reduces damage duration, but it does not reduce the probability of failure, the scope of blast radius, or the existence of the failure itself. These properties are architectural, not operational. They exist before the first trace is ever captured.

The system did not just become more observable. It became more controllable.

Failures were no longer mysteries. They were mechanical problems with mechanical explanations.

APM turned diagnosis from investigation into execution.

This is APM at its highest form:

- Faster isolation of root cause
- Unified technical truth
- Direct protection of revenue

It demonstrates what APM does exceptionally well:

It collapses uncertainty into action.

But it also defines its boundary.

APM accelerates response. It does not decide what risks should exist.

It shortens outages. It does not govern survivability.

It is tactical control.

APM Changed Engineering Culture

Before APM, failures were vague. After APM, failures were attributable.

This created:

- Clear ownership
- Faster recovery
- Accountability

APM turned performance from speculation into evidence.

The Business Value of APM

APM reduces business risk by:

- Shortening outages
- Preventing silent degradation
- Reducing operational cost
- Protecting customer experience

It is one of the highest ROI disciplines in modern IT.

No serious digital business operates without it.

The Horizon of APM

APM begins observing after:

- DNS has resolved
- Routing has converged
- TLS has validated
- Identity trust has succeeded

These assumptions define the outer boundary of APM's authority. APM cannot observe the failure of conditions required for its own existence. When those conditions fail, visibility does not degrade—it disappears. This is why the most catastrophic failures in modern systems leave no APM evidence at all.

If any of those fail, APM observes nothing.

This is not a weakness. It is a boundary.

APM Assumes the World Exists

APM assumes:

- The internet is reachable
- Providers are functioning
- Certificates are valid
- Time is coherent

It observes behavior inside a world that already succeeded.

It cannot observe the failure of that world.

Why APM Cannot Measure Business Survivability

APM measures:

- Efficiency
- Correctness

- Execution health

It does not measure:

- Dependency concentration
- Trust fragility
- Routing exposure
- Certificate collapse

These are not runtime properties. They are structural properties.

And they define business survivability.

APM Is a Microscope

APM is a microscope.

It shows extraordinary detail inside a narrow field of view.

But survivability requires a map, not a microscope.

Microscopes reveal detail. Maps define territory.

APM excels at revealing how failures occur. Software Performance Risk Management exists to govern where failure is allowed to occur at all. One explains incidents. The other constrains exposure. Confusing explanation with governance is how organizations become exceptionally good at diagnosing disasters they never prevented.

It requires seeing:

- What the system depends on
- Who controls those dependencies
- How quickly they can be changed

APM was never meant to answer those questions.

APM is indispensable. It is not sufficient.

It governs execution, not survivability. It accelerates recovery, not prevention. It explains failure, but it does not decide which failures are acceptable. Software Performance Risk Management begins where APM's horizon ends. This at the point where execution is no longer guaranteed and authority must exist before systems are allowed to operate at scale.

4

RUM: Truth After Damage

Before RUM, performance engineering was inward-looking.

APM told teams what the application was doing. Infrastructure monitoring told teams what the servers were doing. Neither told teams what customers were experiencing.

This created a dangerous gap:

- Systems were "healthy"
- Dashboards were green
- Customers were frustrated

RUM was created to close that gap.

Real User Monitoring provides a specific class of truth: consequence. It records what users experienced after systems succeeded in executing. This truth is not partial, inferred, or synthetic. It is human and factual. But it is also temporal. It exists only after damage has already occurred.

It was the first discipline to say: *The only performance that matters is the performance users experience.*

RUM as the Voice of the Customer

RUM instruments the browser or client, not the server.

It measures:

- Page load times
- Rendering delays
- Interaction latency
- Script failures
- Network variability

It captures reality, not theory.

It is the closest thing engineering has to customer testimony.

All Performance Is Personal

All performance is personal.

It always has been. It always will be.

Frustration does not start when a dashboard turns red. It starts at the moment a human being wants to act and cannot.

You click. Nothing happens. You click again. Still nothing. You move the mouse, scroll, tap, refresh. The system is alive, but it is not listening.

That moment is universal. Everyone reading this has lived it in the last thirty days. Many of us in the last thirty minutes.

You feel it before you think it:

- A tightening in your chest
- A sharp breath
- A flash of anger
- A loss of trust

You are not debugging. You are being dismissed.

These reactions are not anecdotal.

They are behavioral evidence of trust erosion. In governance terms, they represent unrecorded loss: abandonment, disengagement, and silent churn. RUM is the only discipline that captures this class of loss directly, without translation, aggregation, or abstraction.

This is what Real User Monitoring truly measures: not latency, not timing, not technical delay. It measures *disrespect*. The system is telling you that your time is less important than its internal confusion.

A survey conducted in the United Kingdom found that slow or unresponsive software costs organizations just under thirty minutes per day, per employee, in lost productivity.[1] Across a workforce, those minutes compound into billions of pounds annually.

This data measures the cost of delay, not the cost of fragility. It quantifies time lost after systems respond slowly, but it does not capture the risk of systems failing entirely. RUM excels at measuring degradation. It is largely silent when systems never load at all.

But that number only measures time. It does not measure emotion. It does not measure erosion. It does not measure resignation.

It does not measure the moment a user quietly gives up. It does not measure the moment an employee stops caring.

Slow software is not merely inefficient. It is demoralizing.

For internal employees, the damage is personal and persistent. They are held accountable for outcomes that depend on systems they cannot control. The system is slow, the report is late, the process fails, and the human being bears the consequence. Over time, this breeds disengagement. Then attrition. Then cultural decay.

For customers, the story is even simpler.

---

[1] "Technology issues identified as a major contributor to the UK's productivity gap – could cost UK PLC £35bn," PR Newswire UK, https://www.prnewswire.co.uk/news-releases/technology-issues-identified-as-a-major-contributor-to-the-uks-productivity-gap-could-cost-uk-plc-35-631234113.html

Two moments of frustration in a single session nearly guarantees abandonment. And when they leave, most do not return.

They do not write angry emails. They do not complain. They do not explain.

They leave. Quietly. Permanently.

And no monitoring system ever sees them go.

This is why Real User Monitoring is beautiful.

It is the only discipline that allows you to stand where the user stands. It does not ask, *"Is the system fast?"* It asks, *"Was the experience humane?"*

By measuring the time between the first user interaction and `domInteractive`[2], RUM captures the distance between intention and fulfillment. It shows you exactly where trust was tested. It shows you where impatience was born. It shows you where respect was lost.

But RUM also carries a quiet sadness.

It is observational. It is post-impact. It tells you what hurt, not why the wound exists.

It shows:

- Where frustration surfaced
- Where hesitation clustered
- Where abandonment became inevitable

But it struggles to show:

- Which architectural decision created the pain
- Which dependency made it unavoidable
- Which pattern will repeat tomorrow

RUM is a mirror. It reflects suffering with honesty. It does not prescribe the cure.

And yet, it is indispensable.

---

[2]domInteractive is the time when as web interface becomes responsive to user input

Because before performance is engineering, before it is economics, before it is architecture, before it is governance,

performance is personal.

It is the memory of rage-clicking a mouse. It is the sigh before closing a tab. It is the decision not to return. It is the moment you stop trusting a system that asked for your attention and did not honor it.

Software Performance Risk Management does not replace RUM. It stands on its shoulders.

RUM tells you *where the human broke.*

What RUM Does That APM Cannot

APM measures execution inside controlled environments. RUM measures experience in uncontrolled environments.

RUM sees:

- Slow consumer networks
- Mobile device constraints
- Browser rendering costs
- Third-party script failures

APM cannot see any of these.

RUM explains why systems that look healthy feel broken.

RUM Changed Engineering Priorities

RUM shifted performance from an internal metric to a human experience.

It forced teams to care about:

- Perceived latency
- Visual stability
- Responsiveness
- Reliability under diversity

This was a cultural correction.

Engineering stopped optimizing machines and started optimizing trust.

The Business Value of RUM

RUM connects performance directly to:

- Conversion rates
- Abandonment
- Revenue
- Customer loyalty

It quantifies:

- How much performance costs
- Where experience breaks
- Who is harmed

RUM is one of the most powerful business intelligence tools ever introduced into performance engineering.

RUM Is Truth, Not Prediction

Formal Distinction: Impact vs. Risk

describes harm that has already occurred.

describes the probability, scope, and persistence of harm that could occur. RUM measures impact with exceptional fidelity.

It does not measure probability, blast radius, or recovery feasibility. As a result, RUM cannot govern business risk, no matter how accurate or comprehensive its data becomes.

RUM answers: *What happened to our users?*

It does not answer: *What is about to happen to our business?*

Truth is retrospective. Risk is prospective.

RUM Begins After Trust Has Succeeded

RUM executes only if:

- DNS resolved
- TLS validated
- Routing succeeded
- The application loaded

If any of these fail, RUM is silent.

The most catastrophic failures never appear in RUM.

RUM silence during catastrophic failure is often misinterpreted as a monitoring gap.

It is not. It is evidence that the system never reached a state where user experience could exist. Failures of trust, routing, identity, or dependency prevent RUM from executing at all. These failures are not invisible. They exist outside RUM's jurisdiction.

RUM Measures Impact, Not Exposure

RUM shows:

- Who was harmed
- How badly
- Where experience degraded

It does not show:

- Why the system was fragile
- How concentrated dependencies are
- Whether the same harm will recur tomorrow

Why RUM Cannot Govern Business Risk

Business risk lives in:

- Probability
- Blast radius
- Recovery time

RUM measures none of these.

It measures consequence.

Consequence is not risk.

RUM Is a Witness

RUM is not a judge. It is not a predictor. It is not a governor.

It is a witness.

Witnesses are essential to justice. But justice also requires prevention.

RUM tells us who was harmed, how badly, and where trust was lost. It does not, and cannot, govern whether the conditions that caused that harm should be allowed to exist again. Software Performance Risk Management does not replace RUM. It gives meaning to what RUM witnesses by governing the risks RUM can only observe after the damage is done.

5

Logs: Engineering Truth, After the Fact

Logs are the oldest and most fundamental form of observability.

Before APM. Before RUM. Before distributed tracing.

There were logs.

They were created for one reason: *To record what actually happened.*

Logs are not opinions. They are not aggregates. They are not inferred behavior.

They are events.

They are the closest thing software has to a factual record of reality.

Logs provide historical truth.

They record what occurred after execution began and after failure unfolded. This truth is precise, durable, and irreplaceable—but it is also temporal. Logs cannot exist until events have already happened. They are evidence of reality, not authority over it.

33

Logs as the Foundation of All Diagnostic Systems

Every modern observability platform is built on logs.

Whether the interface is:

- Splunk
- Sumo Logic
- Dynatrace
- Datadog
- Elastic

The truth source is the same.

Logs are the raw feed of engineering reality.

Metrics summarize behavior. Traces describe flow. Logs record fact.

All higher-order observability depends on them.

Logs Are the Court Transcript of Software

Logs function like a court transcript.

They answer:

- What occurred
- When it occurred
- Where it occurred
- Who executed it

They are admissible evidence inside engineering organizations.

In legal systems, transcripts establish facts.

They do not prevent crimes. They do not determine which actions were permitted. They do not define acceptable risk. Logs function the same way. They establish what occurred with precision. They do not govern what should have been allowed to occur in the first place.

Without logs:

- Outages become speculation
- Root cause becomes narrative
- Accountability disappears

Logs are the backbone of truth.

What Logs Do Exceptionally Well

Logs excel at:

- Root cause reconstruction
- Sequence analysis
- Failure correlation
- Behavioral auditing
- Compliance evidence

They allow engineers to say: *This happened. Then this happened. Then the system failed.*

No other signal is as precise.

Logs Enable Every Recovery Playbook

Every incident response process depends on logs:

- Triage
- Mitigation
- Root cause
- Prevention

Logs are how systems explain themselves after collapse.

Logs are unmatched for explaining failure.

They are ineffective at preventing it. No amount of additional logging can reveal dependency concentration, trust fragility, or recovery infeasibility before failure occurs. Logs scale insight backward in time, not forward.

Logs Are Unmatched for Diagnostic Truth

APM interprets. RUM observes. Logs state.

Logs do not suggest what might have happened. They declare what did happen.

That makes them irreplaceable.

Hero Story: The Day the Checkout Slowed Without Touching the Code

The company was a Top-50 eCommerce platform. Traffic was stable. Marketing was stable. No major code changes had been made to the checkout system.

And yet something was wrong.

Sales order processing was slowing. Checkout abandonment was rising. Revenue was slipping.

NewRelic showed it clearly: transaction times were degrading. But there was no corresponding change in checkout code. No recent deployment explained the behavior.

From the outside, it looked like a mystery.

From the inside, it felt worse:

"We are getting slower without changing anything."

That is the most dangerous failure mode in software.

The answer was not in the application. It was in the logs.

Someone reviewing Splunk noticed something extraordinary. One system stood out violently against all others: the credit card gateway servers. Their log volume was orders of magnitude larger than any other service in the environment.

Not slightly larger. Not gradually larger. Explosively larger.

The question became simple:

"Why is this one system writing so much?"

That question led directly to the root cause.

A new version of the gateway software had recently been promoted to production. Functional tests passed. Cards authorized correctly. Behavior was correct.

But one small change had slipped through.

A developer had changed the logging level to **DEBUG** during troubleshooting and never restored it.

The system used Log4j. Logging was asynchronous. Everyone assumed it was safe.

It was not.

Asynchronous logging does not mean free logging.

It means the application hands work to the operating system. And the operating system always decides what runs first.

In this case:

- Debug logging exploded message volume
- Log buffers saturated
- Disk I/O dominated CPU scheduling
- Kernel activity (ring 0) starved application execution (ring 3)
- The gateway slowed
- Authorization slowed
- Checkout slowed
- Customers abandoned
- Revenue dropped

No exception was thrown. No error appeared. The system was "healthy."

It was simply suffocating under its own logging.

When the log level was corrected, the recovery was immediate.

- CPU normalized
- Disk pressure vanished
- Gateway latency returned to baseline
- Checkout speed recovered
- Abandonment dropped
- Revenue stabilized

One configuration flag had been costing millions.

This is why logs are sacred.

They are not:

- Decoration
- Debug residue
- Developer convenience

They are the forensic truth of system behavior.

Neither APM nor performance testing alone could have revealed this:

- APM showed symptoms
- Performance testing could not reproduce provider-scale constraints
- Logs revealed the physics of failure

Splunk did not merely observe the problem. It explained it.

This is logging at its highest form.

This failure was diagnosable because the system continued to operate long enough to emit evidence.

Logging revealed the physics of collapse only because execution persisted. When failures prevent execution entirely, such as when DNS fails, trust collapses, or routing fractures, no log volume exists to analyze. Logs require a functioning world to speak.

Not as text. Not as noise. But as structural truth.

Logs revealed:

- The real workload
- The real bottleneck
- The real cause
- The real fix

Without logs, this failure would have been:

- Guessed at
- Debated
- Misattributed
- Prolonged

Instead, it was solved.

But the boundary matters.

Logs tell you what happened. They do not tell you what will destroy you.

They diagnose. They do not govern.

They expose failure. They do not prevent catastrophe.

They are history. Now the narrowing begins.

Logs Are Perfectly Historical

Logs exist only after events occur.

They cannot:

- Predict failure
- Measure exposure
- Quantify blast radius
- Estimate recovery constraints

They describe consequence, not probability.

Truth is not foresight.

Logs Cannot See Structural Risk

Logs do not record:

- Dependency concentration
- Provider fragility
- Trust chain collapse potential
- DNS survivability
- Certificate hierarchy exposure

Risk lives in architecture. Logs live in behavior.

The Problem of Time in Logs

Logs depend on timestamps.

But timestamps are only as accurate as the clocks that produce them.

In modern systems:

- Virtual machines drift
- Containers inherit flawed clocks
- Hypervisors resynchronize unpredictably
- NTP adjustments introduce discontinuities

Two logs can be perfectly accurate locally and still be globally wrong.

When Time Lies, Causality Breaks

When causality cannot be established, accountability collapses.

In regulated environments, incorrect timestamps compromise breach timelines, invalidate forensic reconstruction, and undermine contractual responsibility. Time integrity is not an operational detail. It is a legal dependency. Systems that cannot produce coherent time cannot be governed, audited, or defended.

When clocks drift:

- Cause appears after effect
- Recovery appears before failure
- Chains of responsibility become ambiguous

This turns: *Engineering truth into engineering interpretation.*

Time Integrity Is a Legal Problem

Logs are increasingly used as legal artifacts.

They determine:

- Breach timelines
- Regulatory responsibility
- Contractual fault

If timestamps are wrong, evidence is compromised.

Clock integrity is not an engineering detail. It is a legal dependency.

Logs Multiply Data, Not Certainty

More logs create:

- More storage
- More queries
- More cost

They do not create:

- More predictability
- More survivability

Why Logs Cannot Govern Business Risk

Logs are disqualified from governing business risk by design.

They measure consequence, not probability. They reconstruct damage, not exposure. They explain what failed, not what was likely to fail. Treating logs as a risk governance mechanism confuses historical accuracy with foresight—and replaces prevention with narrative.

Logs explain: *What failed.*

They cannot explain: *How likely failure is.*

Business risk lives in:

- Probability
- Blast radius
- Recovery constraints

Logs measure none of these.

Logs Are the Record, Not the Shield

Logs are not prevention. They are memory.

Logs preserve truth. They do not preserve continuity.

They are memory, not protection. Evidence, not authority. When organizations rely on logs to manage survivability, they choose explanation over prevention. Software Performance Risk Management begins where logs end, at the point where truth is no longer sufficient and governance must exist before failure occurs.

6

OpenTelemetry: The Apex of Observability

Can you imagine a marketplace of incompatible observability truths? This led directly to the creation of OpenTelemetry.

Before OpenTelemetry, every vendor defined:

- Its own agents

- Its own data formats

- Its own instrumentation model

- Its own lock-in

Telemetry was not portable. Tracing was not standardized. Metrics were vendor dialects.

Engineering visibility was powerful, but politically fractured.

The Cost of Fragmentation

Engineering teams spent more time:

- Maintaining agents
- Translating telemetry
- Migrating vendors

Than improving systems.

Observability became a tax, not a tool.

Fragmentation did more than slow engineers.

It fractured authority. When execution truth is scattered across incompatible systems, no single version of reality can be governed. OpenTelemetry restored observability as shared infrastructure, making execution truth a stable asset rather than a vendor-controlled artifact.

Why OpenTelemetry Had to Exist

OpenTelemetry represents the apex of observability not because it sees everything, but because it sees everything that can be seen *during execution*. It unifies traces, metrics, logs, and context propagation into a single execution narrative. This is not a tooling milestone. It is a boundary. Once execution is fully observable, remaining failures cannot be blamed on lack of visibility.

OpenTelemetry was not a technical convenience.

It was a governance correction.

It restored:

- Ownership to engineers
- Portability to telemetry
- Neutrality to observability

It unified:

- Traces
- Metrics
- Logs
- Context propagation

Into a single execution language.

The Visibility Revolution

With OpenTelemetry, organizations finally gained:

- End-to-end causality
- Portable diagnostics
- Vendor independence
- Architectural clarity

For the first time in history, execution truth became: *A stable, shared asset.*

The Moment Visibility Became Complete

OpenTelemetry marks the moment observability became complete.

Not perfect. But whole.

No other instrumentation breakthrough remains to be invented.

The execution layer is now fully visible.

Fully visible does not mean fully governable.

Visibility describes what happens when systems execute correctly. It does not govern whether execution should be permitted, whether dependencies are survivable, or whether trust assumptions are defensible. Completeness of visibility exposes—not resolves—these questions.

The Boundary That Even OpenTelemetry Cannot Cross

OpenTelemetry can only observe what executes.

If code never runs, telemetry never exists.

No signal is emitted when:

- DNS fails
- TLS collapses
- Routing fractures
- Trust is revoked
- Time invalidates authentication

These are the highest-impact failures in modern computing.

OpenTelemetry is bounded by execution.

If execution does not begin, no telemetry exists. This is not a limitation of instrumentation. It is a property of reality. Observability cannot record the failure of conditions required for its own existence.

And they are invisible to all execution telemetry.

Perfect Visibility Exposed Structural Blindness

When observability was imperfect, failures could be blamed on instrumentation gaps.

When observability became unified and standardized, something became undeniable:

Systems still failed catastrophically.

This blindness is not observational.

It is structural. Perfect execution visibility reveals where authority is missing. When systems fail despite complete telemetry, the failure lies not in instrumentation, but in governance. Survivability was never assigned an owner.

Not because execution was misunderstood. But because survivability was never modeled.

OpenTelemetry and SPRM Are Parallel Evolutions

OpenTelemetry and SPRM were born from the same realization:

Complexity crossed human intuition.

OpenTelemetry responded by governing execution truth. SPRM responds by governing business exposure.

One answers: *What happened?*

The other answers: *What could happen?*

Engineering and Business Finally Speak the Same Language

With OpenTelemetry, engineers gained clarity. With SPRM, executives gain foresight.

Together they create:

- Technical truth
- Business resilience
- Survivability governance

Two Disciplines, One Direction

OpenTelemetry governs execution truth.

Software Performance Risk Management governs survivability authority. One ensures that what happens is visible. The other determines what is allowed to happen. They are complementary precisely because their jurisdictions do not overlap.

OpenTelemetry perfects how we observe systems. SPRM perfects how we protect organizations.

They are siblings:

- Born of maturity
- Driven by complexity
- Focused on survival

One is the nervous system. The other is the immune system.

Hero Story: When Observability Finally Became Whole

Before OpenTelemetry, most organizations believed they had observability. In reality, they had fragments.

Metrics lived in one system. Traces lived in another. Logs lived somewhere else. Teams owned different tools. Vendors owned different formats.

What was missing was not data. It was continuity.

A public example of OpenTelemetry's impact comes from an engineering team operating a large GraphQL service in a complex, multi-language, distributed environment.[1]

Before adopting OpenTelemetry, the organization struggled with a familiar problem: each system could be observed, but no system could be understood end-to-end. Telemetry existed, but it was trapped inside tool boundaries.

When a request crossed service, language, or platform lines, visibility fractured. Root cause analysis became reconstruction. Diagnosis became interpretation.

Observability was present. Truth was not.

---

[1] "End-User Q&A Series: Using OTel with GraphQL," OpenTelemetry Blog, February 13, 2023, `https://opentelemetry.io/blog/2023/end-user-q-and-a-01/`

By standardizing on OpenTelemetry instrumentation, the team unified their telemetry pipeline across all services, regardless of language runtime or observability backend. Traces were no longer proprietary artifacts owned by a single vendor. They became a shared, portable, architectural signal.

For the first time, engineers could see a request travel from edge to database and back again without losing context.

One example revealed the power of this unification: OpenTelemetry exposed production traces containing three to four thousand spans. These massive traces uncovered deeply nested GraphQL resolver paths and complex downstream service dependencies that had previously been invisible. Without OpenTelemetry, these execution paths simply did not exist in any single system's view of reality.

This was not incremental improvement. It was a change in physics.

OpenTelemetry transformed observability from:

"We can see parts of the system."

Into:

"We can see the system."

For the engineering organization, this meant:

- Unified tracing across heterogeneous stacks
- Elimination of vendor-specific telemetry silos
- True end-to-end request visibility
- Drastically improved root cause isolation

Observability stopped being a collection of tools and became a shared architectural substrate.

This is OpenTelemetry at its apex.

This unification explains execution with unprecedented clarity.

It does not decide whether the architectural complexity it reveals is survivable. OpenTelemetry exposes structural risk; it does not govern it. Visibility without authority still permits collapse.

Not because it is another monitoring solution. But because it is not a solution at all.

It is a standard.

It represents the moment when observability matured from proprietary implementations into common infrastructure, much like TCP/IP matured networking and HTTP matured application communication.

OpenTelemetry did not make observability better. It made it complete.

And that is precisely why Software Performance Risk Management becomes necessary.

When observability is fragmented, failures can be blamed on blind spots. When observability is complete, blind spots disappear.

At that moment, a new question emerges:

"Now that we can see everything, why are we still exposed?"

OpenTelemetry reveals:

- Execution paths
- Dependency chains
- Failure propagation
- Structural fragility

But it does not decide:

- Which risks are acceptable
- Which dependencies are survivable
- Which architectures are defensible

It shows truth. It does not govern consequence.

This is the sibling relationship between OpenTelemetry and SPRM.

OpenTelemetry perfects visibility. SPRM governs survivability.

One answers:

"What is happening?"

The other answers:

"What must never be allowed to happen?"

Together, they form the complete discipline of modern systems stewardship.

7

FinOps Proved the Pattern

This chapter exists as a pause.

It is intentionally separate from OpenTelemetry. Not because the subjects are unrelated, but because they operate in different domains of consequence.

OpenTelemetry solved execution visibility. FinOps solved financial visibility.

Together, they prove a deeper truth:

When technical systems reach sufficient complexity, business governance must follow.

FinOps Was Not About Saving Money

FinOps is often mischaracterized as a cost-cutting initiative.

It is not.

FinOps is not a best practice. It is a precedent.

It demonstrates what happens when technical systems reach a scale where informal oversight fails. Engineering decisions created financial exposure that no existing role was accountable for. The response was not better tooling or more discipline—it was governance. A new authority layer emerged because the system demanded it.

FinOps exists because cloud systems made financial exposure invisible.

Before FinOps:

- Engineers made infrastructure decisions
- Finance saw invoices
- No one could connect cause to consequence

FinOps created a translation layer between:

- Technical architecture
- Financial accountability

It was not about spending less. It was about understanding what spending meant.

FinOps Was a Governance Correction

FinOps emerged when three conditions converged:

1. Engineers controlled decisions that created material financial exposure

2. Finance owned outcomes without visibility into causes

3. No existing discipline could translate between the two

Governance did not appear because of inefficiency. It appeared because accountability was misaligned. FinOps corrected that misalignment by making cost a governed technical property.

FinOps emerged because:

- Engineering decisions created financial exposure

- Finance could not reason about systems

- Engineers could not reason about financial impact

So a new discipline was created.

Not to control engineers. But to connect engineering reality to business consequence.

FinOps Did Not Replace Engineering

FinOps did not replace:

- Architecture

- Cloud engineering

- Platform design

It surrounded them.

FinOps did not remove authority from engineers.

It constrained it. Engineers retained decision-making power, but those decisions became legible, reviewable, and accountable at the business level. Governance did not override engineering judgment. It bounded it.

It added:

- Cost visibility

- Financial accountability

- Decision traceability

FinOps made technical systems financially legible.

The Structural Parallel to SPRM

The structural parallel is exact:

- FinOps governs financial exposure created by technical systems
- SPRM governs survivability exposure created by technical systems

In both cases, engineers create exposure through architectural decisions. In both cases, executives bear responsibility for outcomes they cannot directly influence. In both cases, metrics existed long before governance did, and proved insufficient.

FinOps governs: *How expensive it is to operate.*

SPRM governs: *How dangerous it is to fail.*

Both exist because:

- Technical complexity exceeded informal oversight
- Metrics alone became insufficient
- Business impact became opaque

Cost Is a Continuous Risk

Cost accumulates gradually.

It is a continuous risk:

- Overspend
- Waste
- Inefficiency

Failure risk is different.

It is discontinuous.

Continuous risks can be optimized. Discontinuous risks must be governed.

Cost accumulates gradually and tolerates correction. Failure occurs suddenly and tolerates none. This difference explains why optimization disciplines are insufficient for survivability. You cannot tune your way out of collapse. You can only decide, before it happens, how much exposure is acceptable.

It is:

- Sudden
- Binary
- Existential

This is why FinOps and SPRM are siblings, not substitutes.

FinOps Made Engineering Speak Business

FinOps introduced a new expectation:

Engineering decisions must be explainable in financial terms.

SPRM introduces the next expectation:

Engineering decisions must be explainable in survivability terms.

Why FinOps Makes SPRM Inevitable

Once an organization accepts that financial exposure created by technical systems must be governed, it has already conceded the principle that technical systems create business risk beyond engineering control. At that point, rejecting survivability governance is not a philosophical disagreement, it is an inconsistency.

Once an organization accepts:

- Financial governance of technical systems

It becomes impossible to argue against:

- Risk governance of technical systems

FinOps opened the door.

SPRM must now walk through it.

Two Axes of Accountability

FinOps created:

- Cost accountability

SPRM creates:

- Survivability accountability

Together they form:

- Operational legitimacy
- Executive governance
- Architectural responsibility

This Chapter Exists to Set the Precedent

FinOps proves that:

When engineering matures, business governance must follow.

OpenTelemetry proved execution could be unified. FinOps proved cost could be governed. SPRM proves survival must be governed.

This chapter is not a detour. It is evidence.

OpenTelemetry marks the point at which execution can no longer hide.

FinOps did not succeed because it was persuasive. It succeeded because the system demanded accountability.

Software Performance Risk Management exists for the same reason. Complexity has outgrown intuition. Exposure has outgrown dashboards. Survivability has outgrown informal ownership. FinOps proved that governance emerges when systems require it.

SPRM is not a question of adoption. It is a question of timing.

CHART OF FRAGILE WATERS

— A Navigation Guide to Software Performance Risk —

OPEN SEA OF SOFTWARE SYSTEMS

- SINGLE DNS PROVIDER
- DEFAULT CART ARCHITECTURE
- UNBOUNDED THIRD PARTY SCRIPTS
- REEFS OF HIDDEN FOUNDATIONS

HERE BE CASCADING FAILURES

- TRAFFIC SPIKES
- VENDOR OUTAGES
- ROUTING FAILURES
- CERTIFICATE EXPIRY
- CLOCK DRIFT

STORMY SEAS

ABANDON HOPE, YE WHO IGNORE RISK SURFACES

Compass:
- R — PROBABILITY
- BLAST RADIUS
- W
- E
- RECOVERY
- S

- PERFORMANCE TESTING
- APM
- RUM
- LOGS
- SYNTHETIC MONITORING

SPRM MATURITY PASSAGE

SAFE HARBOR (SPRM)

NO MONITORING ALONE CAN SAVE YOU.

Legend

- **Trade Routes:** *Traditional Performance Tools*
- **Reefs:** *Hidden Foundations*
- **Storm Zones:** *Unbounded Risk*
- **Compass:** *Risk Equation*
- **Channel:** *SPRM Maturity Passage*
- **Harbor:** *Governed Systems*

Not all who wander are lost.
Some are mapping danger for others.

Part II

Defining Software Performance Risk Management

This part establishes Software Performance Risk Management as a formal discipline.

Not as a product. Not as a framework. Not as a methodology.

As a governance domain.

Part I demonstrated that modern performance engineering is mature, powerful, and complete – and that its completeness revealed a structural blind spot. Part II defines what fills that void.

Performance engineering governs how systems run. Software Performance Risk Management governs whether organizations survive when they do not.

This is the beachhead.

From this point forward, SPRM is no longer a reaction to the limits of existing tools. It becomes its own system of thought, language, and responsibility.

Here we define:

• What "performance risk" actually means
• How it differs from performance optimization
• How it is measured
• How it is governed

This is where performance stops being an engineering concern and becomes a business survival property.

No Answers

I was in my truck, driving. That's always been my thinking space. The place where hard problems get taken apart and rebuilt over and over until something finally gives. The road gives me time. Silence. A rhythm that lets complicated systems simplify themselves.

The phone rang. It was Brian Bernknopf , a colleague at QA Consultants. Facebook had just gone offline.[1] A BGP event. Global. Immediate. Catastrophic. Brian asked a simple question: How could we have found something like this ahead of time to help clients?

I opened my mouth to answer and realized I had nothing. Three decades in this industry. Operating systems. Databases. Networks. Test tools. Performance engineering. Consulting. Architecture. At scale. In production. In failure. And in that moment, I had no answer.

Not a weak answer. Not a partial answer. Nothing.

And then it happened again.[2] Different company. Different failure. Same question. And still I had nothing.

Then came the CDN outages[3] Tens of thousands of sites disappearing at once. Commerce stopped. Governments went dark. Communication failed. Once again, I was asked how we could have helped clients see it coming.

And once again, I had nothing.

That was the moment the ground shifted.

Not because the outages were surprising. We had always known systems were fragile. But because we had built an entire profession around explaining failure after it happened, and almost nothing around governing the risk of it happening at all.

We had become very good at measurement. Very good at diagnostics. Very good at dashboards. Very good at telling stories about what broke.

And somehow, we had convinced ourselves that meant we were responsible.

We weren't.

We were observers in a system that desperately needed owners.

---

[1] https://en.wikipedia.org/wiki/2021_Facebook_outage
[2] https://www.thousandeyes.com/blog/internet-report-episode-47
https://www.thousandeyes.com/blog/rogers-outage-analysis-july-8-2022
[3] Fastly, June 8 2021. Akamai Edge DNS, July 22 2021. AWS US East, Dec 7 & 10 2021. Cloudflare, June 22 2022. Rogers Communication, July 2022. Google Cloud, August 9 2022

So I started questioning the way we ask questions about performance. Not how fast something is. Not how it behaves under load. Not how quickly we can recover. But why entire classes of failure remained invisible to every discipline I respected and every tool I had helped build.

There had to be a better way to see the risks that were clearly there, waiting quietly in architecture, dependency, trust, and assumption.

That question followed me everywhere. Behind the wheel. In the seat of a rowing machine. Walking the roads and trails near my home in the foothills of South Carolina.

It wasn't a single insight. It was a long reconciliation between facts I already knew but had never allowed to touch.

We were not failing at performance. We were failing at responsibility.

And once I saw that, I could not unsee it.

What eventually emerged was not a new metric or a new tool. It was a governance model. A way to treat software performance risk as something that belongs at the same level as financial risk, operational risk, and reputational risk.

That realization changed everything.

Because from that point forward, performance was no longer a technical discipline to be optimized. It was a responsibility to be owned.

Human operators exercising finite authority within systems that exceed individual control.

8

The Human Drivers of Software Performance Risk Management

This chapter does not describe human weakness. It describes predictable failure modes of human systems operating under scale, incentive pressure, and incomplete authority.

Human behavior is not the problem. Human behavior is the constraint.

In complex systems, humans operate with bounded attention, asymmetric incentives, and limited authority. When systems exceed those bounds, failure becomes deterministic. Software Performance Risk Management exists because no amount of training or individual excellence can reliably compensate for that mismatch.

Software is no longer just infrastructure. It is society.

During COVID, software became the only interface between people and essential outcomes: food, healthcare, income, education, safety. Systems that had been designed for convenience were suddenly carrying consequence.

People were not browsing. They were trying to survive.

Unemployment portals collapsed. Health systems timed out. Aid platforms failed silently.

There were no alternate channels. No human overrides. Just absence.

That absence was not neutral. It was frightening.

This was not a performance problem. It was a human one.

We asked systems built for optimization to perform governance. They were not designed for that role.

COVID did not create fragile systems. It revealed them.

And the same fragility exists everywhere.

In commerce. In education. In identity. In trust.

We pause our lives for software more often than we admit.

A ticket sale freezes. A registration portal fails. A payment times out.

These are not outages. They are interruptions in human narrative.

Moments where intention meets silence.

And those moments are recorded—not in logs or dashboards—but in public memory.

Social platforms have become the world's largest archive of performance failure. Not measured in milliseconds, but in disappointment. Not expressed in charts, but in consequence.

Performance risk is no longer merely operational.

It shapes:

- who receives care
- who accesses education
- who gets aid
- who trusts institutions

When society depends on software, software becomes governance.

And governance without resilience becomes accidental cruelty.

This is why Software Performance Risk Management must exist.

Not because systems are slow. Not because observability is incomplete.

But because performance failure now carries human cost.

Traditional performance disciplines ask:

How fast is the system?

SPRM asks:

Who suffers when it fails?

They ask:

How long was the outage?

SPRM asks:

What moment was stolen?

Anything that carries human consequence must be governed, not merely optimized.

Failure Mode 1 — Incentive Blindness

Incentive Blindness occurs when locally rational optimization produces globally unbounded risk because the optimizing actor does not own survivability outcomes.

- Arises under delivery, availability, or cost pressure
- Increases exposure without triggering feedback
- Persists even with good intent
- Cannot be corrected by awareness

Incentive blindness cannot be solved by better incentives alone. Any system that allows exposure without accountability will reproduce it. Only governance that constrains survivability exposure can arrest this failure mode.

Failure Mode 2 — Delegation Decay

Delegation Decay occurs when responsibility fragments faster than authority can follow, creating ownership vacuums where survivability risk accumulates.

- "That's owned by another team"
- Survivability assumptions embedded in interfaces
- No single escalation path for cross-domain failure
- Incidents attributed to "complexity"

Delegation decay is not a coordination failure. It is a governance failure caused by authority fragmentation.

Failure Mode 3 — Dashboard Sedation

Dashboard Sedation occurs when visibility suppresses intervention by signaling normalcy while exposure grows outside the visible surface.

Dashboards do not fail by omission. They fail by reassurance.

This failure mode cannot be corrected by adding metrics. It emerges whenever dashboards are treated as control mechanisms rather than observation tools.

Why These Failures Persist

These failure modes appear in competent, well-intentioned organizations. They persist because they are properties of scale, not skill.

As systems grow, exposure grows faster than human judgment. When survivability depends on vigilance or heroics, failure is not a possibility—it is an eventual certainty.

These failures cannot be trained away or observed away.

Software Performance Risk Management exists to govern what humans cannot reliably control: survivability exposure that accumulates before incidents occur.

9

What Is Software Performance Risk Management

Software Performance Risk Management does not promise prevention, prediction, or perfection. It exists to determine whether survivability risk is explicit, bounded, and owned before failure occurs.

SPRM does not eliminate uncertainty. It eliminates unmanaged exposure.

Software Performance Risk Management (SPRM) is a governance discipline. It exists to assign accountability for survivability exposure in software systems before incidents occur.

Software Performance Risk Management is the discipline responsible for:

- identifying survivability exposure created by software systems,

- bounding that exposure to governable limits,

- and assigning decision authority for accepting, mitigating, or rejecting that exposure.

Modern systems are no longer bounded by a single team, runtime, or authority. They are:

- composed of external dependencies,
- governed by third parties,
- bound to trust fabrics,
- exposed to global failure domains.

Performance *risk* has therefore ceased to be a purely technical property. It has become a business survivability property.

> **Axiom**
>
> **Jurisdictional Definition**
>
> Software Performance Risk Management is the only discipline accountable for **pre-incident survivability exposure**.
>
> Survivability exposure includes architectural, dependency, trust, and recovery conditions that determine whether a system can continue to operate when stressed, degraded, or partially failed.

When performance collapses today, it is no longer an engineering inconvenience. It is a revenue event, a reputational event, and often a legal or regulatory event.

SPRM exists because performance now directly shapes business continuity.

Performance Failures Are Business Events

A modern performance failure is indistinguishable from a business crisis.

When performance fails:

- revenue pipelines stop,
- customer trust is damaged,
- legal exposure increases,
- market confidence erodes.

Yet most organizations still evaluate performance through engineering artifacts:

- dashboards,
- SLAs,
- error rates,
- test results.

These describe system behavior. They do not describe survivability exposure.

SPRM introduces the language required to make performance failures governable by executives, not just diagnosable by engineers.

SPRM as Business Risk Governance

SPRM treats performance risk as business risk: not cost optimization, but loss exposure.

It governs whether loss is *bounded* by:

- constrained blast radius,
- feasible recovery,
- diversified dependencies,
- survivable trust reliance.

This mirrors how other governance disciplines operate: they do not predict catastrophe; they constrain the conditions under which catastrophe becomes ungovernable.

SPRM as Technical Survivability Governance

SPRM does not replace engineering. It governs what engineering alone cannot.

Engineering asks: *Does the system work?*

SPRM asks: *Can the organization survive when it doesn't?*

It evaluates:

- dependency concentration,
- recovery feasibility,
- trust fragility,
- control boundaries.

These are architectural properties, not performance metrics.

SPRM as Structural Classification

SPRM accelerates recognition of recurring failure structures.

Traditional performance practice discovers systemic risk:

- during outages,
- after failures,
- through long incident cycles.

SPRM identifies structural recurrence across systems:

- repeating dependency collapse patterns,
- common concentration surfaces,
- recurring trust and recovery failure modes.

It turns anecdote into governed categories, reducing time to executive comprehension and preventive decision.

Why Existing Models Cannot Become SPRM

Performance testing measures behavior under cooperation. APM measures behavior under execution. RUM measures behavior under experience. Logs measure behavior after failure.

None of them are designed to govern:

- unbounded blast radius,
- infeasible recovery,
- brittle trust dependence.

This is a categorical limitation, not an implementation flaw.

On the Nature of Risk Models

The models introduced in SPRM are not predictive engines. They do not forecast outages, calculate likelihoods, or produce statistically precise outcomes.

Their purpose is decision constraint. They exist to force explicit reasoning about consequence, containment, and accountability — not to simulate the future.

A model that enables better decisions without predicting events is not incomplete. It is fit for governance.

Failure Condition

SPRM becomes necessary when software systems can fail in ways that:

- are not testable,
- are not observable during execution,
- and cannot be mitigated by operational response.

At that point, survivability becomes a governance problem rather than an engineering one.

What SPRM Explicitly Does Not Do

- SPRM does not test software.
- SPRM does not instrument software.
- SPRM does not monitor runtime behavior.
- SPRM does not respond to incidents.
- SPRM does not optimize performance.
- SPRM does not replace SRE, APM, RUM, security, or FinOps.

SPRM governs whether the conditions under which these disciplines operate are survivable.

Authority Model

SPRM does not execute survivability controls. It governs whether survivability risks are acceptable.

Execution remains distributed across engineering, architecture, operations, and security—but authority over whether those efforts are sufficient must be centralized. Governance is not monopoly. It is accountability.

SPRM requires decision authority. Survivability exposure cannot be governed without the ability to constrain architectural choices, dependency adoption, and risk acceptance. Where this authority does not exist, SPRM cannot function.

SPRM accountability is violated when survivability exposure is accepted implicitly, inherited unknowingly, or deferred without an owning decision.

- Performance Engineering → optimizes execution
- Observability → explains execution
- SRE → stabilizes execution
- Incident Response → restores execution
- Security → protects execution
- FinOps → governs cost of execution
- **SPRM → governs survivability of execution**

SPRM does not compete with these disciplines. It constrains the risk envelope within which they operate.

Limitation

SPRM does not prevent all failures. It does not eliminate uncertainty. It does not guarantee uptime.

It ensures that survivability risk is explicit, bounded, and owned before failure occurs.

Closing

Software Performance Risk Management exists because survivability risk cannot be delegated, inferred, or ignored without consequence. Where survivability exposure exists without authority, failure is inevitable. SPRM assigns that authority.

10

The Risk Equation

The SPRM risk equation is not a predictive model. It does not forecast incidents or compute timelines. It exists to determine whether survivability risk is **bounded or unbounded**.

Every mature discipline eventually develops an equation.

Finance models portfolio exposure. Insurance models catastrophe impact. Engineering models load, stress, and failure.

Performance engineering never required a true risk equation because it optimized behavior, not survival.

As long as performance meant speed and efficiency, dashboards were sufficient. Latency charts and throughput curves described success.

Once performance failure became a business-threatening event, a different model was required.

Executives do not manage milliseconds. They manage exposure.

They need to understand:

- whether failure initiation is constrained,
- how large the impact would be if it occurs,
- and whether recovery is feasible before material harm accrues.

Without all three dimensions, no discussion of risk is complete.

Why This Is a Bounding Equation, Not a Predictive Model

Predictive models require stable inputs, repeatable conditions, and historical completeness. Survivability risk in modern software systems satisfies none of these assumptions.

The SPRM risk equation therefore does not attempt prediction. It determines whether risk is constrained within limits that governance can act upon.

This is the difference between forecasting outcomes and governing exposure.

Executive Brief

Most boards speak about risk as something that can be insured, diversified, or transferred. That is often true for financial instruments. It is not true for software.

Software risk is embedded in architecture, dependencies, trust relationships, and recovery design. It is not something that "happens to" an organization. It is something an organization creates—deliberately or not.

The risk equation in this chapter is not mathematics. It is a responsibility construct.

It forces a simple realization: risk is the intersection of failure initiation uncertainty, blast radius, and recovery feasibility. Every system already has this equation. The only question is whether leadership is conscious of it.

Probability is not luck. Blast radius is not coincidence. Recovery is not optimism.

When a system fails, the organization does not discover its risk posture. It reveals it.

This is where software performance becomes a fiduciary concern.

Why Existing Metrics Cannot Express Risk

Traditional performance metrics answer operational questions:

- How fast is the system?
- How much load can it handle?
- How often does it error?

These are behavioral truths. They are not survivability truths.

Risk requires reasoning about:

- uncertainty of failure initiation,
- structural vulnerability,
- magnitude and duration of consequence.

A system can be fast, stable, and efficient—and still be existentially fragile.

No existing performance metric can express that condition.

The Risk Equation

All business risk can be expressed using three dimensions:

Risk = Initiation Uncertainty × Blast Radius × Recovery Feasibility

This structure is not new. It appears across finance, insurance, disaster modeling, and safety engineering.

What is new is its application to software performance.

Performance failures are no longer local defects. They are systemic business threats.

They therefore require the same analytical framing as other catastrophic risks.

Initiation Uncertainty: Can Failure Be Bounded?

Initiation uncertainty in SPRM is not a calculated probability. It reflects whether failure initiation can be structurally constrained.

Failure initiation becomes ungovernable when driven by:

- dependency monoculture,
- opaque third-party control planes,
- deep trust chains,
- emergent interactions outside organizational authority.

A system that depends on a single DNS provider, CDN, or trust authority is initiation-fragile regardless of normal performance.

Initiation uncertainty is architectural. It is not load-dependent.

> **Axiom**
>
> Survivability risk is governable only when:
> - failure initiation is bounded,
> - blast radius is constrained,
> - and recovery is feasible.
>
> If any condition is unbounded, survivability risk exceeds the authority of engineering disciplines and requires governance intervention.

Blast Radius: How Much Breaks When It Breaks?

Formal Definition

In SPRM, blast radius is not a metaphor. It is a structural property of a system that describes the maximum scope of organizational harm resulting from the failure of a dependency.

It is determined by architecture, coupling, authority concentration, and irreversibility — not by runtime conditions.

Because blast radius is largely fixed at design time, it is governable. And because it is governable, failure to bound it is an act of omission, not misfortune.

Blast radius is the scope of impact once failure occurs.

Two systems may fail equally often. One disrupts a feature. The other threatens the enterprise.

Blast radius is driven by:

- architectural centralization,
- identity and payment coupling,
- communication concentration,
- dependency convergence.

Blast radius is not a technical metric. It is a business impact boundary.

> **Axiom**
>
> **Governability Test**
>
> If blast radius is unbounded, the system is ungovernable.

Recovery Feasibility: Can Damage Be Contained?

Recovery in SPRM is not about speed. It is about feasibility.

If recovery requires conditions that cannot be restored—revoked trust, unavailable routing, or inaccessible third-party control—then recovery time is irrelevant. Survivability has already failed.

Recovery feasibility is constrained by:

- exit complexity,
- control over naming and routing,
- certificate and trust reissuance,
- organizational decision latency.

Recovery is organizational before it is technical.

Why the Equation Is Multiplicative

The equation is multiplicative because these dimensions amplify one another.

A failure that is unlikely, contained, and recoverable is tolerable. A failure that is uncertain, widespread, and unrecoverable is existential.

Optimization of a single dimension is insufficient. Survivability emerges only when all three are constrained.

How the Risk Equation Changes Decision Making

Without the equation:

- risk is debated,
- responsibility is diffused,
- decisions are emotional.

With the equation:

- tradeoffs become explicit,
- exposure becomes governable,
- accountability becomes unavoidable.

Limitation

The SPRM risk equation does not rank systems numerically. It does not compare applications by score.

It determines whether survivability risk can be governed at all.

When survivability risk is unbounded, no amount of testing, monitoring, or response can substitute for authority. The equation does not predict collapse. It identifies when collapse cannot be prevented without governance.

11

Hidden Foundations

Hidden foundations are not abstractions. They are real dependencies, control planes, and assumptions that determine whether software systems can continue to operate under stress.

They are considered "hidden" not because they are unknown, but because they fall outside the authority of teams responsible for delivery and operation.

Executives rarely fail because they make poor operational decisions. They fail because structural commitments were made silently—dependency choices, architectural shortcuts, and trust assumptions that later determine whether impact is contained or catastrophic.

> **Axiom**
>
> **Hidden foundations** are the infrastructural, organizational, and trust dependencies upon which software systems rely, but which are not explicitly governed as survivability risks.

This chapter surfaces those commitments—not to assign blame, but to make responsibility visible. Once foundations are seen, they can be governed. Until then, survivability remains accidental.

Boundary Clarification: Structure and Human Response

Software Performance Risk Management does not claim that architecture alone determines outcomes, nor that human judgment, operational excellence, or incident response are irrelevant. Organizations routinely avert catastrophe through skill, improvisation, and experience.

SPRM asserts something narrower and more precise: when survivability depends on sustained human intervention, vigilance, or heroics, failure is no longer a possibility—it is an eventual certainty. Human response can delay consequences, but it cannot reliably govern unbounded structural exposure at scale.

SPRM therefore governs what humans cannot consistently compensate for: architectural, dependency, and trust conditions that determine whether failure remains containable once attention, time, or coordination are exhausted.

Modern software systems appear to be defined by code. In reality, they are defined by foundations that exist beneath code and outside organizational control.

These foundations include:

- Naming and resolution (e.g., DNS)
- Time and synchronization
- Identity and trust anchors
- Routing and reachability
- External control planes (cloud, SaaS, CDN, identity providers)
- Organizational ownership boundaries

This list is intentionally incomplete. SPRM does not require completeness. It requires that survivability-critical foundations be made explicit and owned.

None of these are owned by the application team. Most are barely understood by it.

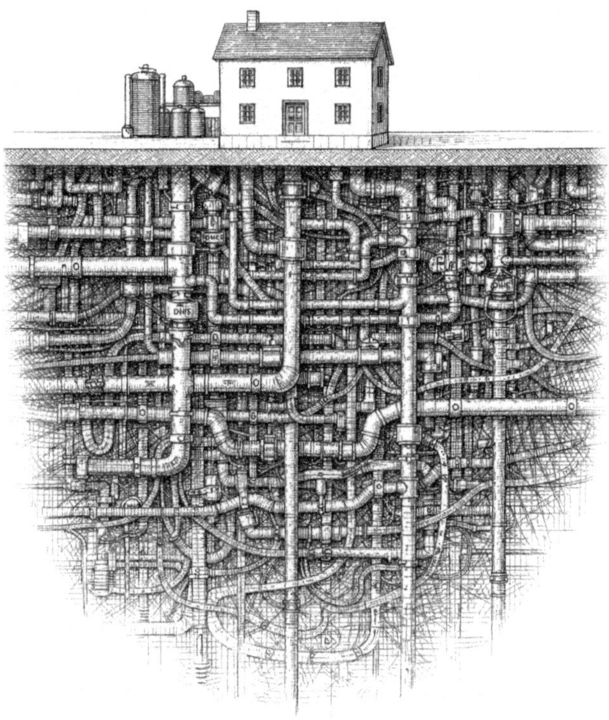

Yet they determine whether the system exists at all.

SPRM does not attempt to surface all dependencies. It governs only those dependencies whose failure would produce unbounded blast radius or infeasible recovery.

Hidden foundations persist because no existing discipline owns their survivability implications:

- Performance testing assumes their availability
- Observability requires their operation
- SRE responds after their failure
- Security protects them without governing dependence on them

As a result, survivability characteristics are inherited implicitly rather than governed explicitly.

Executive Brief

Every organization believes it understands its critical systems. Very few understand what those systems are built upon.

Hidden foundations are the invisible structures of dependency, trust, and architectural assumption that determine whether a system survives stress or collapses under it. They are why two organizations with similar teams and similar technology can experience radically different outcomes under failure.

Executives rarely see these foundations because they do not appear in dashboards. They do not show up in uptime reports. They are not visible in financial statements.

Yet they quietly determine how fragile or resilient an organization truly is.

A single DNS provider. A single CDN. A single authentication authority. A default shopping cart architecture. A third-party script in a browser header. A dependency that cannot be bypassed during failure.

None of these are accidents. They are choices — often made for speed, cost, or simplicity — and almost always made without a survivability discussion.

This chapter is about making those choices visible.

Hidden foundations explain why outages cascade instead of isolate. Why recovery becomes chaotic instead of controlled. Why organizations are surprised by failures that were architecturally inevitable.

This chapter reframes software risk as a structural problem, not an operational one. It asks leadership to stop asking "How do we respond faster?" and start asking "Why does our system have this much power to hurt us?"

Once hidden foundations are exposed, risk becomes governable. Until then, it remains accidental.

Protocol Infrastructure Is the Real Platform

Most organizations believe their platform is:

- Kubernetes
- Cloud providers
- Application frameworks

Those are execution platforms.

The real platform is:

- DNS
- TLS
- BGP
- NTP

These protocols define:

- Whether a system is reachable
- Whether it is trusted
- Whether it is routable
- Whether it is temporally valid

No performance test, APM trace, or RUM measurement exists without them.

They are the bedrock of the internet. And therefore the bedrock of business continuity.

Infrastructure Truth: When BGP Took Facebook Off the Internet

On October 4th, 2021, Facebook disappeared from the Internet.

Not metaphorically. Not partially. Not degraded.

It vanished.

Facebook, Instagram, WhatsApp, Messenger, and Oculus were unreachable worldwide for more than six hours. Billions of users could not connect. Businesses that relied on Facebook's platforms were instantly offline. Advertising revenue stopped. Customer support channels collapsed. Even internal Facebook systems became inaccessible.

This was not an application failure. It was not a data center failure. It was a routing failure.

During routine maintenance, Facebook's network configuration caused its Border Gateway Protocol (BGP) routes to be withdrawn from the global routing table. In simple terms, the Internet no longer knew how to find Facebook.

No DNS record could resolve to it. No packet could reach it. No monitoring system could observe it.

The company did not go down. It was erased.

According to public reporting, Facebook's prefixes were removed from global BGP advertisements, meaning that Internet service providers and backbone networks had no path to route traffic to Facebook's infrastructure.[1]

This is the reality of BGP.

It decides whether your systems are reachable at all. It operates below DNS, below CDNs, and below observability.

If BGP fails, nothing above it matters.

APM has no signal. RUM has no users. Logs have no entries.

From the perspective of application tooling, the system does not degrade. It ceases to exist.

This incident exposes a truth most organizations are unprepared to confront:

Your application does not control its own existence.

---

[1] "2021 Facebook outage," Wikipedia, `https://en.wikipedia.org/wiki/2021_Facebook_outage`

Your survivability is delegated to:

- Routing authorities
- Peering relationships
- Transit providers
- Network configuration correctness
- Protocol stability

And none of those are visible to application observability.

This is why BGP belongs in SPRM.

Not because it fails often, but because when it fails, nothing else matters.

Traditional monitoring assumes reachability. SPRM treats reachability as a risk surface.

Traditional systems ask:

"How fast is my system?"

SPRM asks:

"Can my system be found at all?"

That is the difference between engineering excellence and survivability governance.

Third-Party Dependencies Are Architectural Decisions

Every time a team selects:

- A CDN
- A payment processor
- An identity provider
- A messaging service
- An analytics platform

they are making an architectural decision that affects survivability.

These are not convenience integrations. They are risk allocations.

Each dependency imports:

- Its own failure probability
- Its own blast radius
- Its own recovery constraints

Dependency selection is survivability design.

CDNs: Acceleration and Fragility

CDNs are often treated as performance tools. They are survivability infrastructure.

They centralize:

- traffic delivery
- TLS termination
- DDoS protection
- caching

This makes them powerful. It also makes them catastrophic single points of failure.

When a CDN fails, businesses disappear instantly across geography.

The decision to use a CDN is also the decision to trust its failure modes and recovery constraints.

Infrastructure Truth: Shared Dependency Collapse (Fastly, 2021)

On June 8th, 2021, a global outage at Fastly caused widespread disruption across thousands of dependent sites. A latent software bug was triggered by a customer configuration change, impacting large portions of Fastly's edge network.[2]

---

[2]Fastly, "Summary of June 8 Outage," `https://www.fastly.com/blog/summary-of-june-8-outage`

The lesson for SPRM is not the bug. It is the blast radius created by shared dependency concentration.

Different industries. Different architectures. One provider. One failure domain.

If a single third-party control plane can erase your public existence, that dependency is a hidden foundation and must be governed as survivability exposure.

Surface Dependencies Are Invisible Load Multipliers

Modern sites load:

- dozens of JavaScript libraries
- third-party pixels
- advertising networks
- A/B testing frameworks
- payment widgets

Each dependency adds:

- latency
- failure probability
- trust exposure

Surface dependencies are architectural risk disguised as convenience.

The Hidden Cost of Architectural Decisions

Architectural decisions carry cost in three forms:

- operational cost
- performance cost
- survivability cost

The third is almost never evaluated.

Examples:

- centralized authentication services
- single-region databases
- monolithic APIs
- hard-coded provider dependencies

They optimize simplicity. They concentrate blast radius.

Architectural Bombs

Some design decisions function as delayed explosives.

They appear safe until scale arrives.

Examples:

- synchronous global locks
- chatty service meshes
- single-writer architectures
- centralized state stores
- cross-region consensus dependencies

These create:

- performance ceilings
- recovery impossibility
- exponential cost curves

They are architectural bombs.

SPRM exists to detect them before detonation.

The Architectural Bomb: The Default Cart

Some architectural failures are loud. They crash immediately. They announce themselves early.

The most dangerous ones are silent.

They work perfectly—right up until the moment they destroy you.

The default shopping cart is one of those bombs.

Across the internet, millions of eCommerce sites carry it. It passes functional testing. It works under normal load. It even works under moderate success.

And then, at the precise moment of maximum opportunity, it explodes.

There is a rule of thumb in performance engineering that has existed for decades:

> If more than 50% of users will need a resource, pre-allocate it.
> If fewer than 50% will need it, allocate it on demand.

Pre-allocation means:

- memory reserved
- database rows created
- sessions opened
- locks taken
- resources held whether they are used or not

In most eCommerce systems, fewer than 20% of sessions ever use a shopping cart.

Most visitors browse. They compare. They leave.

The cart is an exception state.

And yet, millions of sites allocate it by default.

Now imagine a successful recording artist. Millions of fans. An active online store.

One day, they announce:

> "In five minutes, I'm dropping 100 signed, 180-gram vinyl records. First come, first served."

Three hundred thousand fans arrive within seconds.

Now watch the difference between two architectures.

In a just-in-time model:

- Only the first 100 fans who attempt to add the item to a cart get one
- 100 carts are created
- 100 inventory locks are applied
- The rest are told "sold out"
- The system remains calm
- The database remains stable
- The site survives

In a default cart model:

- All 300,000 sessions are issued carts
- All 300,000 carts are persisted
- All 300,000 carts touch storage
- All 300,000 carts consume memory
- All 300,000 carts require cleanup later
- All 300,000 carts hammer the database within seconds

The site must now be engineered not for purchasing, but for waste.

The architecture is forced to support:

- massive over-allocation
- session cleanup storms
- lock contention
- disk saturation
- cache collapse

The bomb detonates.

When it goes off, the pattern is always the same:

- cart creation slows
- database queues grow
- locks accumulate
- thread pools exhaust
- memory fragments

- new sessions stall
- the site locks

No one can buy. No one can browse. The store is "up" but dead.

Recovery is not immediate because the system must drain. Carts time out slowly. Resources release in waves.

Functionally, the system was correct. Architecturally, it was doomed.

This pattern is not limited to shopping carts. It appears wherever systems:

- allocate resources before intent is known
- hold resources longer than necessary
- treat convenience as architecture
- confuse functional correctness with survivability

Too early. Too long. Too optimistic.

That is the signature of an architectural bomb.

Architectural bombs are not bugs. They are decisions.

And decisions are governance problems, not engineering problems.

That is why survivability must be designed.

Hidden Foundations Determine Recovery Feasibility

Recovery depends on whether:

- DNS can be changed quickly
- certificates can be replaced under duress
- routing can be altered
- providers can be exited

If these foundations are rigid, recovery becomes a negotiation with systems you do not control.

Hidden foundations define how trapped an organization is.

Hidden Foundations Are Rarely Documented

Architecture diagrams show:

- services
- databases
- APIs

They rarely show:

- trust chains
- DNS provider diversity
- certificate authority dependency
- routing concentration
- third-party blast radius

What is undocumented becomes unmanaged.

Hidden Foundations Are the Core of SPRM

Before SPRM, organizations optimized what they could see.

SPRM governs what they inherit and assume:

- foundational dependencies,
- concentrated authority,
- survivability constraints that cannot be fixed during an incident.

Hidden foundations are where software stops being code and becomes continuity.

Governance Artifact: Foundation Inventory

SPRM requires an explicit inventory of survivability-critical foundations.

This inventory does not describe implementation detail. It records:

- the foundation
- the owning entity
- the failure mode
- the blast radius if it fails
- whether recovery is feasible without external consent

Authority Implication

A foundation without an owner is a survivability liability.

SPRM requires that survivability authority be assigned for each foundation. Where authority cannot be assigned—because control lies outside the organization—dependency risk must be explicitly accepted or constrained.

Closing

Hidden foundations decide whether failure is containable.

When foundations are ungoverned, survivability becomes accidental and recovery becomes a negotiation with systems you do not control.

Software Performance Risk Management exists to make these foundations explicit, bounded, and owned before failure turns them into a crisis.

12

From Metrics to Exposure

Modern technology organizations are rich in metrics.

They measure latency, error rates, throughput, saturation, and availability. These metrics describe system behavior. They are engineering truth.

But business risk does not live in behavior. It lives in vulnerability.

Exposure is the condition that allows failure to propagate.

It exists before incidents occur, before probabilities are estimated, and before systems are stressed. Software Performance Risk Management treats exposure as a *structural* property of systems, not a statistical one.

Exposure

Exposure is the extent to which a system's survivability depends on conditions it does not control.

Exposure exists regardless of whether failure is imminent. It is a property of architecture, dependency, trust, and recovery—not of execution.

A system can exhibit perfect operational metrics and still carry catastrophic exposure.

Metrics may indicate: *The system is healthy.*

Exposure asks: *What happens if it is not?*

SPRM exists because organizations have mastered measurement but failed to govern exposure.

Exposure vs. Probability

Probability describes how often failure may occur. Exposure describes how much damage occurs when it does.

Systems can have low failure probability and catastrophic exposure. SPRM governs the latter.

Classes of Survivability Exposure

Dependency exposure reliance on external services or control planes

Concentration exposure single points of architectural dominance

Trust exposure reliance on identities, certificates, or authorities

Recovery exposure inability to restore required conditions

Organizational exposure unclear ownership or authority

This list is intentionally incomplete. SPRM does not require exhaustive enumeration. It requires that material exposure be visible and owned.

Limitation

SPRM does not attempt to eliminate exposure. It governs whether exposure is acceptable, bounded, and survivable.

Why Exposure Escapes Existing Disciplines

Exposure is not observable during normal operation.

- Performance testing assumes exposure
- Observability reveals impact only after exposure is exercised
- SRE mitigates consequences after activation
- Security protects mechanisms, not dependency reliance

As a result, exposure accumulates silently until failure activates it.

Governance Artifact: Exposure Register

SPRM requires an explicit exposure register.

This register records:

- the exposure
- the affected system
- the dependency or condition involved
- the blast radius if exercised
- recovery feasibility
- the authority responsible for acceptance

An exposure register is not a mitigation plan. It is an accountability instrument.

Authority Implication

Exposure without an owning authority is unmanaged risk.

SPRM requires that exposure be either constrained by design or explicitly accepted by an accountable authority. Silent exposure is a governance failure.

Why Metrics Cannot Govern Risk

Metrics are inherently reactive. They describe what is happening, what has degraded, and what is failing now.

Risk governance requires understanding what *could* happen, how severe the impact would be, and whether recovery is feasible.

Metrics are snapshots. Exposure is topology.

Topology cannot be governed with telemetry alone.

Exposure as a Board-Level Artifact

Exposure is not a dashboard metric. It is a board-level governance artifact.

It informs:

- architectural investment
- vendor selection
- dependency acceptance
- regulatory preparedness
- business continuity planning

SPRM elevates performance risk from operations into governance.

Closing

Exposure determines how systems fail, not whether they fail.

When exposure is ungoverned, survivability is accidental. Software Performance Risk Management exists to ensure that exposure is explicit, bounded, and owned *before* failure makes it visible.

13

Passive Risk Intelligence

Passive Risk Intelligence is the disciplined use of survivability-relevant information that already exists, without installing agents, modifying systems, or altering execution.

It requires:

- no code changes
- no instrumentation
- no runtime dependencies
- no operational permission

Passive Risk Intelligence does not observe behavior. It observes *structure*.

Definition of Passive Risk Intelligence

Passive Risk Intelligence is the analysis of survivability-relevant structure using externally observable, configuration-derived, or protocol-level signals that exist independently of application execution.

It reveals how failure *would propagate*, not how software behaves under normal operation.

What Passive Risk Intelligence Is Not

Passive Risk Intelligence is not:

- monitoring
- observability
- telemetry
- logging
- surveillance
- inspection of application data

It does not explain execution. It constrains survivability assumptions.

Structural Signals Used

Passive Risk Intelligence draws from signals such as:

- DNS topology and resolution authority
- TLS certificates and trust chains
- BGP routing diversity and reachability
- CDN provider concentration
- Email authentication posture
- Public dependency relationships

These signals exist independently of any application. They are structural facts of the internet.

This list is intentionally incomplete. SPRM does not require comprehensive visibility. It requires sufficient structural evidence to bound survivability exposure.

Why Passive Intelligence Is Required

Many survivability failures originate outside application execution.

When survivability depends on external authorities, shared control planes, or trust anchors, runtime telemetry is insufficient to describe risk.

Passive Risk Intelligence exists because these dependencies:

- are assumed rather than governed
- are invisible during healthy operation
- determine blast radius and recovery feasibility

When Passive Intelligence Is the Only Option

Passive Risk Intelligence is required when:

- applications cannot be instrumented
- infrastructure is not directly controlled
- third-party systems must be assessed
- mergers or acquisitions are evaluated
- regulatory or fiduciary risk is examined

In these contexts, agent-based approaches cannot operate.

Why Agents Cannot See Structural Risk

Agents execute only after:

- name resolution succeeds
- routing converges
- trust validates

If any of these fail, agents never run.

Agents observe success conditions. SPRM governs failure conditions.

This is not a tooling limitation. It is a categorical boundary.

What Passive Risk Intelligence Cannot See

Passive Risk Intelligence cannot observe:

- malicious insider intent
- zero-day logic defects
- social engineering attacks
- legal or contractual constraints
- human decision-making under stress

Passive Risk Intelligence does not predict incidents. It bounds survivability assumptions by revealing structural dependence.

Governance Artifact: Passive Intelligence Map

SPRM produces passive intelligence maps that record:

- survivability-critical dependencies
- control plane ownership
- concentration and monoculture
- trust anchor reliance
- implicit recovery assumptions

These maps are not diagnostic tools. They are governance instruments used to accept, constrain, or redesign exposure.

Authority Implication

Passive Risk Intelligence reveals exposure but does not resolve it.

When passive intelligence identifies unbounded survivability exposure, SPRM requires that an accountable authority either constrain that exposure or explicitly accept it.

Passive intelligence without authority produces insight without control.

Closing

Passive Risk Intelligence does not make systems safe.

It makes survivability assumptions explicit.

Software Performance Risk Management relies on passive intelligence not to predict failure, but to determine whether failure would be survivable when it occurs.

If you know the enemy and know yourself, you need not fear the result of a hundred battles.

Sun Tzu
The Art of War

14

Risk Surfaces

Risk does not enter an organization evenly. It enters through specific surfaces—points where dependency, assumption, and control intersect.

These are the locations where localized failure becomes systemic impact, where technical fragility becomes business exposure, and where survivability is decided before incidents occur.

If leadership cannot name these surfaces, leadership cannot govern risk.

Definition: Risk Surface

A **risk surface** is a boundary at which failure can propagate beyond the control of a single component, team, or authority.

Risk surfaces are not abstract categories. They are concrete interfaces where assumptions cross control boundaries.

Risk Surfaces Cut Across Categories

Boards often discuss risk in categories:

- Cyber
- Operational
- Financial
- Regulatory

Risk surfaces cut across all of them.

They are tangible:

- a CDN dependency
- a DNS provider
- a payment gateway
- a browser-side script
- an identity provider
- a default architectural pattern

Each is a doorway through which failure can enter the business.

This chapter is about identifying *where* organizations are structurally vulnerable, not where they have already failed.

Once risk surfaces are visible, risk becomes governable. Until then, resilience is accidental.

The Three Classes of Risk Surfaces

Most organizations treat risk as something that comes from "outside":

- Cloud providers
- CDNs
- Payment processors
- Identity platforms

That is only one of three survivability risk surfaces.

In practice, performance risk accumulates across:

1. External third-party dependencies

2. Internal surface dependencies

3. Unsurfaced architectural performance debt

The first is widely discussed. The second is occasionally acknowledged. The third is almost entirely ignored.

SPRM exists to govern all three.

Risk Surface 1: External Third-Party Dependencies

External dependencies define who controls your availability and recovery.

These include:

- DNS providers
- CDNs
- Certificate authorities
- Payment gateways
- Identity providers
- Messaging platforms

Each introduces:

- independent failure modes
- independent recovery timelines
- independent governance models

Engineering teams often treat these as integrations. SPRM treats them as survivability contracts.

Every external dependency implicitly answers:

- Who controls availability?
- Who controls recovery?
- Who decides when the system can resume operation?

This is not technical trivia. It is operational sovereignty.

The Script in the Header

No one noticed it at first.

It was one line of JavaScript, added late on a Friday afternoon...

That one line of code was not "monitoring."

It was a transfer of survivability authority.

This is the class of failure SPRM exists to surface.

It is not:

- a bug
- a performance regression
- a capacity issue

It is a trust decision embedded in architecture.

Boundary Clarification

This example does not argue that third-party scripts are categorically unsafe.

It demonstrates that any dependency placed on the critical execution path becomes a risk surface whose survivability characteristics must be governed.

Risk Surface 2: Internal Surface Dependencies

Surface dependencies are components loaded into the execution path that are not under full organizational control.

Examples include:

- analytics platforms
- advertising networks
- chat widgets
- A/B testing frameworks
- tag managers

Each:

- executes in the critical path

- shares the brand

- shares the blast radius

They silently expand exposure while remaining largely invisible to traditional tooling.

Risk Surface 3: Unsurfaced Architectural Performance Debt

This is the most dangerous surface because it is structurally invisible.

Teams test execution paths and optimize what exists. They rarely ask whether the architecture itself is survivable.

Examples include:

- centralized state stores

- global synchronous locks

- serial dependency chains

- single-region authority

These patterns pass tests. They pass metrics. They fail under real stress.

Limitation

SPRM does not claim that all failures originate at risk surfaces.

It governs the surfaces where failure propagation becomes unbounded and recovery becomes infeasible.

Governance Artifact: Risk Surface Map

SPRM produces risk surface maps that record:

- the surface or boundary
- the survivability assumption being made
- the controlling authority
- the blast radius if the assumption fails
- recovery feasibility

These maps are not operational diagrams. They are governance instruments.

Authority Implication

A risk surface without authority is an unbounded blast radius.

SPRM requires that survivability authority be assigned at each risk surface. Where authority cannot be assigned, exposure must be explicitly accepted at the executive level.

Closing

Risk surfaces determine how far failure can propagate.

When they are ungoverned, blast radius becomes accidental.

Software Performance Risk Management exists to identify these surfaces and ensure that survivability exposure across them is explicit, bounded, and owned.

15

Trust as Infrastructure

Trust is not a cultural attribute. In software systems, trust is a dependency.

Every organization operates on a set of implicit assumptions: that names will resolve, certificates will validate, routes will converge, identities will authenticate, and messages will deliver. These assumptions are not abstract beliefs. They are infrastructural conditions that determine whether software can execute at all.

When these conditions fail, systems do not degrade gracefully. They stop.

Trust, once broken, is extraordinarily expensive to restore—not because of sentiment, but because the underlying infrastructure that produces credibility, availability, and legitimacy has failed.

Trust as a Structural Dependency

Boards often treat trust as a matter of brand, communications, or compliance. In practice, trust is manufactured by infrastructure.

Naming systems, routing protocols, certificate authorities, identity providers, email delivery chains, and third-party platforms silently mediate every transaction. These systems rarely appear in financial reporting, yet they determine whether an organization is perceived as reliable or fragile at moments that define reputation.

Definition: Trust in SPRM

Trust, in Software Performance Risk Management, is the reliance of a system on external authorities to validate identity, integrity, or permission as a prerequisite for execution.

Trust is therefore not an assurance. It is a survivability dependency.

What SPRM Does Not Do With Trust

SPRM does not:

- authenticate users
- authorize actions
- issue or manage credentials
- design identity architectures
- replace security or identity engineering

SPRM governs the organization's *dependence* on trust systems. It does not operate those systems.

The Operational Culture of Trust

Most organizations operate under an unspoken assumption:

"If a major provider offers a service, it must be reliable."

This assumption shapes architecture, vendor selection, and risk acceptance. Yet many trust decisions are made without survivability modeling, exit planning, or recovery guarantees.

Trust is often granted by convenience rather than governance.

The Myth of SaaS Guarantees

SaaS providers commonly advertise availability and redundancy. What they rarely guarantee is survivability under stress: recovery speed, dependency collapse behavior, or blast radius containment.

An application can be technically "up" and still be unusable.

Uptime is not survivability. Availability is not recoverability.

SPRM treats vendor assurances as claims until their failure characteristics are structurally understood.

Trust Without Exit Is Dependency Concentration

Every trust decision must be paired with an exit path.

If an organization cannot:

- change DNS providers rapidly
- re-issue certificates under duress
- re-route traffic independently
- replace an identity provider

then trust has become captivity.

Trust without exit is not trust. It is dependency concentration.

DNS: Trust in Existence

DNS is the trust fabric of reachability. When DNS fails, systems do not degrade, they vanish.

Historical failures such as the Dyn outage (2016) and Facebook's DNS misconfiguration (2021) were not performance incidents. They were trust collapses.

CDN: Trust in Delivery

CDNs terminate encryption, absorb attack traffic, and mediate global routing. When they fail, the blast radius is planetary.

These events are not capacity failures. They are survivability failures rooted in trust concentration.

BGP: Trust in Direction

BGP determines where traffic flows and depends on mutual trust between network operators. When that trust fails, traffic is hijacked or black-holed.

These failures are invisible to application telemetry. They are infrastructural in nature.

Email: Trust in Identity

Email is the trust fabric of communication. Failures in email trust enable fraud, impersonation, and reputational collapse without touching application servers.

Trust failures here are business failures, not technical ones.

Limitation

SPRM does not attempt to make trust systems more secure.

It governs whether organizational reliance on those systems creates unbounded blast radius or infeasible recovery.

Governance Artifact: Trust Dependency Register

SPRM requires an explicit register of trust dependencies.

This register records:

- the trust system
- the operating authority
- what execution depends on it
- blast radius if trust is unavailable
- recovery feasibility if trust is degraded or revoked

This register is not a security inventory. It is a survivability accountability instrument.

Authority Implication

Trust dependencies without survivability authority create hard-stop failure modes.

SPRM requires that organizational reliance on trust systems be explicitly governed. Where trust dependencies cannot be constrained, their risk must be accepted at the executive level.

Trust, But Verify

SPRM provides verification not through promises or contracts, but through structural analysis.

It verifies:

- who controls resolution
- who controls routing
- who controls trust issuance
- who controls recovery

Trust is no longer assumed. It is governed.

Closing

Trust determines whether software is permitted to operate.

When trust systems are ungoverned dependencies, survivability becomes fragile. Software Performance Risk Management exists to ensure that organizational reliance on trust is explicit, bounded, and owned before failure makes it visible.

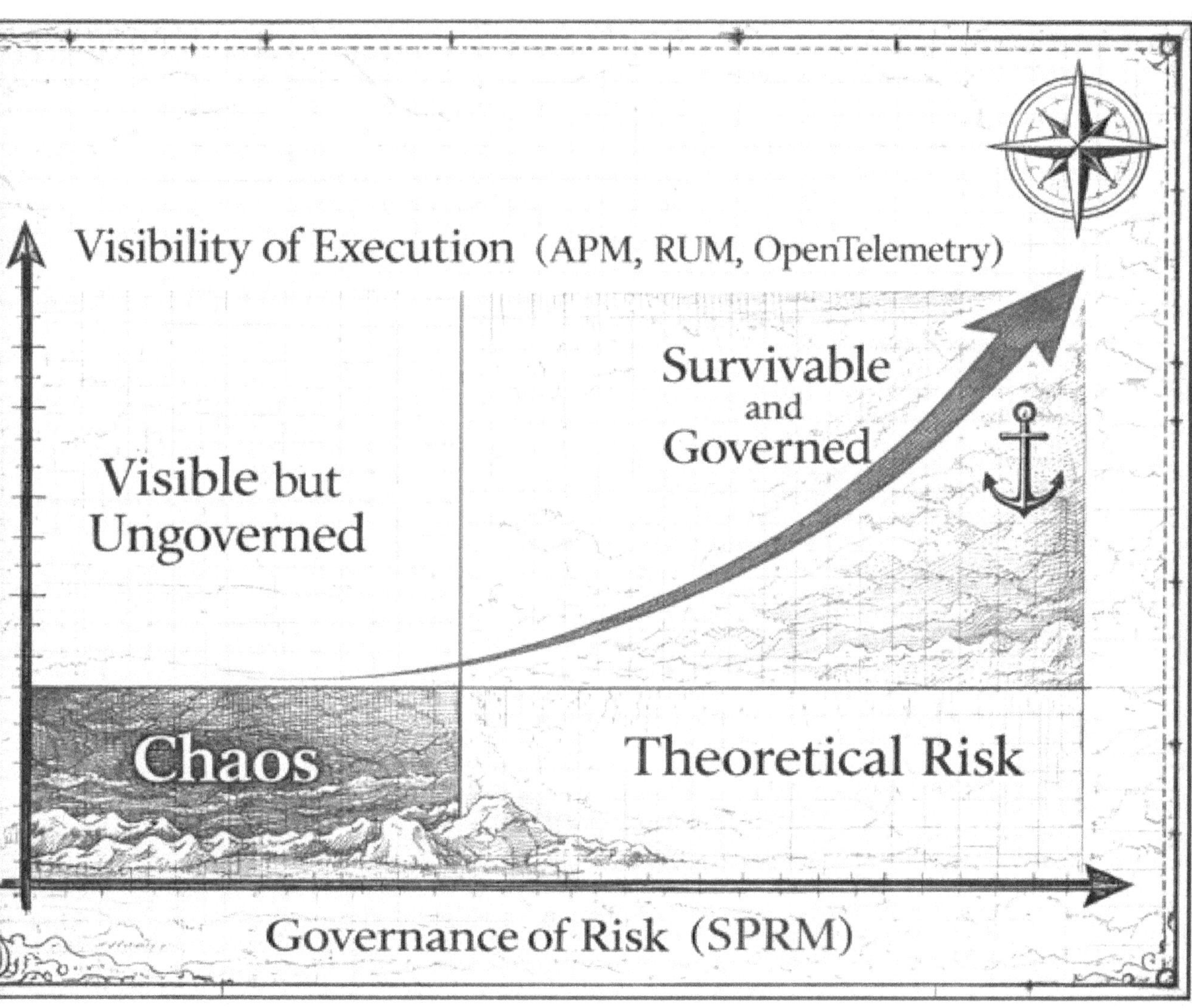
Visibility of Execution (APM, RUM, OpenTelemetry)
Survivable
and
Governed
Visible but
Ungoverned
Chaos
Theoretical Risk
Governance of Risk (SPRM)

Part III

Building the Safe Harbor

Part I established why performance tools alone cannot protect a business. Part II defined how performance risk actually operates.

Part III addresses the only remaining question:

Under what conditions can an organization legitimately claim survivability?

This part does not describe a transformation. It defines minimum conditions.

Organizations do not leap from dashboards to governance. They cross thresholds.

Most organizations stop at **dashboards. Risk lives above them.**

The SPRM Maturity Model is not aspirational. It is disqualifying.

Each level represents a survivability threshold. If the conditions of a level are not met, survivability at that level cannot be claimed.

Progression through these levels requires:

- Explicit assignment of responsibility
- Increased visibility of structural exposure
- Transfer of decision authority over survivability

As organizations advance, survivability authority necessarily moves away from local optimization and toward explicit governance. This transfer of authority is not optional. Survivability exposure cannot be governed without it.

This part of the book provides a structured path for establishing a safe harbor without disrupting execution-focused engineering practices and without relying on cultural reform as a risk control.

Survivability is not assumed. It is built, constrained, and owned.

16

Stewardship

There is a moment in every profession when the work stops being about skill and starts being about stewardship.

A pilot can fly long before they understand that lives depend on their judgment. A doctor can practice medicine long before they accept that uncertainty is their constant companion. A builder can raise structures long before they realize people will trust them to stand.

Software has reached that moment.

For years, performance was something we optimized. Risk was something we mitigated. Failures were something we explained.

But at scale, explanation is not enough. Optimization is not enough. Mitigation is not enough.

Software Performance Risk Management does not require centralized control, nor does it prescribe a single organizational model. Authority in SPRM does not mean hierarchy; it means **clear ownership of survivability decisions.**

Distributed teams may retain full autonomy over design and delivery while survivability authority is explicitly assigned at the points where architectural, dependency, or trust decisions create unbounded exposure.

SPRM constrains *risk acceptance*, not *innovation*. Where authority is absent, autonomy becomes accidental risk transfer rather than empowered decision-making.

At scale, visibility becomes responsibility.

The arc of Software Performance Risk Management begins with ownership.

Not ownership of tools. Not ownership of dashboards. Ownership of consequence.

When you see a survivability risk and choose not to act, you have made a decision. When you fail to measure survivability exposure, you have accepted it. When you design a dependency without understanding its blast radius, you have authorized it.

SPRM begins at the moment an organization acknowledges that performance risk is not emergent behavior, but authored outcome.

You wrote this system. You chose its shape. You chose what it trusts. You chose what it concentrates. You chose what it ignores.

And therefore, you choose whether it is survivable.

Pre-SPRM Condition (Level 0)

Before stewardship, there is denial.

Organizations at Level 0 may have:

- Extensive dashboards
- Mature performance engineering
- Robust incident response

But they lack:

- Explicit ownership of survivability exposure
- Visibility into structural dependency risk
- Authority to constrain architectural risk

Disqualifier: If survivability exposure exists without an accountable owner, the organization is not practicing SPRM at any level.

SPRM does not begin with tooling. It begins with responsibility.

The Stewardship Threshold

The maturity model that follows is not a ladder of sophistication. It is a sequence of disqualifying thresholds.

Crossing the first threshold requires one decision:

Survivability exposure must be visible, owned, and consciously accepted.

Until this decision is made, no claim of resilience, reliability, or preparedness is valid.

The Maturity Thresholds

The SPRM maturity levels represent increasing willingness to accept accountability:

- **Observed** — Survivability exposure is acknowledged as real.
- **Measured** — Structural exposure is made visible.
- **Risk Aware** — Exposure is evaluated before failure.
- **Managed** — Intervention is permitted.
- **Governed** — Accountability is explicit and enforced.

Each transition represents not more data, but more authority.

Authority Implication

At the point stewardship begins, survivability decisions can no longer be made implicitly.

Authority Requirement: If no individual or body has the authority to constrain architecture, dependencies, or trust assumptions on survivability grounds, stewardship has not been achieved.

Advisory roles are insufficient. Awareness without authority is theater.

From Excuse to Ownership

Before SPRM, when systems failed, the narrative was familiar:

"We didn't know." "We couldn't see it." "It came out of nowhere."

After stewardship, the narrative changes:

"We understood the exposure." "We evaluated the tradeoff." "We accepted the risk."

This is not perfection. It is accountability.

And accountability is the minimum standard for systems that now govern access to money, education, healthcare, identity, culture, and safety.

Governance Artifact: Survivability Declaration

The entry artifact for stewardship is simple and explicit:

This organization acknowledges that survivability exposure exists, that it carries human and business consequence, and that responsibility for governing that exposure is assigned.

Without this declaration, maturity claims are invalid.

The Quiet Reward

Stewardship does not produce heroism. It produces calm.

It is a launch day where no one is holding their breath. A board meeting that is not defensive. An on-call rotation that feels uneventful. A system that behaves as expected under stress.

It is not applause. It is confidence.

And confidence is what civilization demands from systems that now run lives.

SPRM is not a framework for control. It is a declaration of stewardship.

*Stewardship begins when responsibility
for consequence is consiously eccepted.*

17

From Observed to Measured: Escaping the Dashboard Ceiling

Dashboards describe motion. Risk describes fate.

Most organizations believe they are measuring performance risk because they have dashboards. In reality, they are observing execution after failure conditions already exist.

This chapter defines the transition from **Observed** to **Measured**, the first enforceable step into Software Performance Risk Management.

This transition is not cultural. It is procedural. It is bounded. And it can be completed without agents, code changes, or organizational disruption.

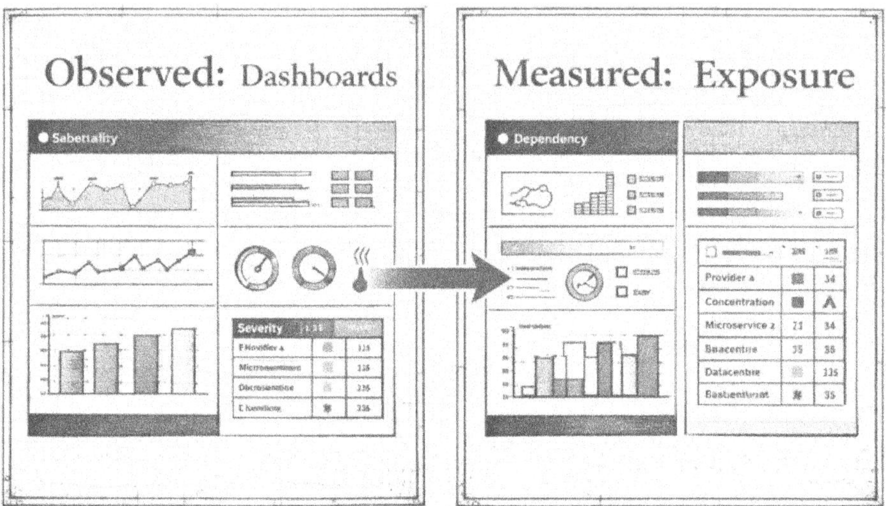

Figure 17.1: Dashboards reveal activity. Exposure reveals danger. SPRM begins when observation becomes measurement.

What "Observed" Really Means

An organization at the **Observed** level has:

- Dashboards
- Metrics
- Alerts
- Incident response procedures

It knows:

- When failures occur
- How often systems degrade
- Which component failed last

It does **not** know:

- Which dependencies control survivability
- Where blast radius concentrates
- Which failures are existential versus tolerable
- Whether recovery is feasible without external consent

Observed means:

> "We can see failures after they happen."

It does **not** mean:

> "We understand our risk."

Dashboards are necessary. They are not sufficient.

What "Measured" Actually Means

Measured is not about precision. It is about **constraint**.

At the **Measured** level, an organization can state—before failure:

- Which dependencies are survivability-critical
- Which failures would be existential
- Which providers control recovery
- Where blast radius is concentrated

Measured means:

> "We can rank our survivability threats."

This is the moment risk becomes:

- Comparable
- Discussable
- Governable

Measured Disqualifier

If an organization cannot identify at least one dependency whose failure would cause existential business harm *before* an incident occurs, it is not Measured — regardless of tooling, dashboards, or documentation.

Measurement that does not surface existential exposure is documentation, not governance.

Observability as a Data Partner

Observability systems are not competitors to SPRM. They are evidence providers.

Observability:

- Collects execution facts
- Records system behavior
- Confirms failure characteristics

SPRM:

- Analyzes structural exposure
- Determines survivability
- Governs risk acceptance

Observability is the Joe Friday of the system:

"Just the facts."

SPRM identifies the criminal.

Critical boundary: Observability can explain how a system failed. It cannot determine whether failure was survivable.

That determination requires governance, not telemetry.

The Transition: From Observation to Measurement

The transition from Observed to Measured consists of five bounded steps. Completing them does not make an organization safe. It makes its danger visible.

Step 1: Define the Risk Surface

Specific:

- Enumerate externally reachable domains

- Identify DNS providers

- Identify CDN providers

- Identify certificate authorities

- Identify email trust posture

Artifact:

- One authoritative inventory

- One dependency graph

Time-bound:

- 5 business days

Step 2: Establish Passive Data Feeds

Specific:

- DNS observation

- TLS certificate inspection

- BGP route visibility

- Surface dependency enumeration

Constraint:

- No agents
- No production changes

Time-bound:

- 10 business days

Step 3: Build the First Exposure Register

Specific:

- One row per survivability-relevant dependency
- Provider identity
- Concentration level
- Replacement feasibility

This is not an asset list. Assets describe existence. Exposure describes consequence.

Time-bound:

- 5 business days

Step 4: Assign Qualitative Exposure Levels

Specific:

- Low / Medium / High exposure
- Based on blast radius and recovery feasibility

Measured outcome:

- Every survivability-relevant dependency classified

Time-bound:

- 3 business days

Step 5: Produce the First SPRM Report

Specific:

- Top survivability risks
- Single points of existential failure
- Recovery constraints

Audience:

- Executives, not engineers

Time-bound:

- 5 business days

Common Failure Mode: False Measurement

Producing inventories, diagrams, or reports without evaluating blast radius and recovery feasibility does **not** constitute measurement.

It produces documentation. Not governance.

A Concrete Shift

Before Measured:

> "We use multiple CDNs and monitor all of them."

After Measured:

> "Two CDNs are existential. One is recoverable within hours. Two are irrelevant to survivability."

That difference is measurement.

What Changes After This Level

After Measured:

- Risk stops being anecdotal
- Discussions move from opinion to ranking
- Survivability enters architectural vocabulary

The organization crosses a line:

From:

"We respond to outages."

To:

"We understand which outages would destroy us."

This is the first moment Software Performance Risk Management becomes real.

This chapter is the on-ramp. It proves SPRM is executable. It proves it is operational. It proves it is governance-grade.

This is where observation ends. And responsibility begins.

18

From Measured to Risk Aware

Measured organizations can see exposure. Risk Aware organizations understand consequence.

This chapter marks the transition where Software Performance Risk Management stops being an inventory exercise and becomes a discipline of judgment.

Measured maturity answers the question:

What are we structurally exposed to?

Risk Aware maturity answers a harder one:

Which of these exposures actually matter?

This distinction is subtle, but decisive. Many organizations reach Measured maturity and stall. They accumulate registers, dependency graphs, and exposure tiers, yet still fail catastrophically when stress arrives. Not because the data was wrong, but because meaning was never extracted from it.

Risk awareness is the moment exposure becomes interpretable in business terms.

Why Measurement Alone Is Insufficient

Measurement produces visibility. Visibility does not produce prioritization.

At the Measured level, organizations can say:

- What they depend on
- Where single points of failure exist
- Which providers are concentrated

But they cannot yet say:

- Which failures would halt revenue
- Which outages would permanently damage trust
- Which incidents would create regulatory or legal exposure

All exposure is not equal. Treating it as such is itself a risk.

Risk Aware maturity exists because leadership does not govern lists. They govern consequence.

The Shift From Technical Exposure to Business Meaning

Risk awareness begins when technical exposure is translated into organizational harm.

This does not require new tools. It does not require new metrics. It requires a change in framing.

Instead of asking:

How fragile is this dependency?

The organization asks:

What happens to the business if this fails?

This is the moment where SPRM leaves engineering language behind and adopts the vocabulary of executives, boards, and fiduciaries.

The Risk Equation Enters the Conversation

At Risk Aware maturity, the organization can reason about risk using a shared model:

$$\text{Risk} = \text{Probability} \times \text{Blast Radius} \times \text{Recovery Feasibility}$$

This equation does not predict incidents. It frames accountability.

Measured maturity touches probability through exposure analysis. Risk Aware maturity introduces the other two terms:

- **Blast Radius** — how much of the organization is harmed
- **Recovery Feasibility** — whether damage can be undone

Without these, exposure remains abstract. With them, survivability becomes governable.

What Changes at Risk Aware Maturity

An organization that has crossed into Risk Aware maturity can:

- Rank survivability risks by consequence
- Separate nuisance failures from existential ones
- Explain technical risk in plain business language
- Defend prioritization decisions to leadership

Conversations change.

Before:

"Everything is important."

After:

"These three failures would cripple us. The rest are tolerable."

That sentence alone signals Risk Aware maturity.

Authority Implication

Risk awareness introduces an unavoidable governance question:

Who is empowered to act on this knowledge?

Once consequences are understood, inaction becomes a decision. Risk Aware maturity therefore requires at least provisional authority to influence:

- Architectural prioritization
- Dependency selection
- Recovery investment

If insight cannot shape decisions, the organization remains Measured in name only.

Why Blast Radius Comes Next

All organizations believe they understand probability. Few understand blast radius.

Probability feels familiar. Blast radius feels personal.

It forces uncomfortable realizations:

- That a small technical failure can halt an entire business
- That trust collapses faster than systems
- That recovery time is often irrelevant once damage escapes containment

Blast radius is where software performance stops being technical and becomes existential.

Hand-Off to the Next Chapter

This chapter establishes *why* Risk Awareness is required. The next chapter defines *how* consequence is understood.

The concept that enables this transition is blast radius.

Not as metaphor. Not as rhetoric. As a structural property of systems that determines whether failure is survivable or fatal.

Measured organizations know what can fail. Risk Aware organizations know what failure would cost them.

The next chapter introduces blast radius as the missing dimension of software performance risk, and the point at which survivability becomes a business discipline rather than an engineering concern.

19

Introducing Blast Radius

Measured organizations know what they depend on. Risk Aware organizations know what those dependencies can destroy.

This chapter introduces blast radius—the dimension of software performance risk that transforms technical exposure into business consequence.

Blast radius answers a question dashboards never can:

If this fails, how much of the organization is harmed?

Until this question is answered, performance risk remains abstract, negotiable, and routinely underestimated.

Boundary Clarification: Probability and Survivability

Software Performance Risk Management does not reject probability as meaningless. Probability remains relevant for operational planning, capacity management, and expected incident frequency.

SPRM rejects only one practice: allowing low probability to justify unbounded consequence. When blast radius is existential or recovery is infeasible, probability ceases to be a governing variable. Survivability risk is not eliminated by rarity.

SPRM therefore treats probability as insufficient—not irrelevant—when evaluating whether failure consequences exceed the organization's ability to absorb them.

Why Blast Radius Matters More Than Probability

Most organizations obsess over likelihood:

- How often failures occur
- How reliable providers claim to be
- How stable systems appear under test

Probability feels scientific. Blast radius feels uncomfortable.

But in survivability analysis, probability is the least stable variable. Blast radius is largely determined at design time.

A failure that is:

- Rare
- Unpredictable

Can still be catastrophic if its blast radius is unbounded.

Blast radius explains why:

- Small configuration errors erase companies
- Minor provider outages become global incidents
- Systems that "usually work" still destroy trust

Risk awareness begins when organizations stop asking *whether* failure will occur and start asking *what happens when it does.*

What Changes From Measured to Risk Aware

At the Measured level, an organization can say:

- What it depends on
- Where exposure exists
- Which dependencies are concentrated

At the Risk Aware level, the organization can say:

- Which failures halt revenue
- Which failures collapse trust
- Which failures create regulatory or legal exposure

Measured is technical awareness. Risk Aware is business awareness.

This is the first maturity level where executives can meaningfully engage.

Risk Aware maturity enables executive understanding, not executive safety. It allows leaders to see which failures matter—but not yet to prevent them. Until architecture, dependency selection, and recovery paths are actively redesigned, risk remains understood but not constrained.

Risk Aware maturity is therefore a necessary condition for governance, not a sufficient one.

Blast Radius as a Structural Property

Blast radius is not a runtime metric. It is a property of architecture, dependency, and trust.

It is determined by:

- Centralization of control
- Coupling of business functions
- Concentration of authority
- Irreversibility of failure

A dependency's blast radius does not change under load. It changes only when architecture changes.

This is why blast radius belongs in SPRM.

Operationalizing Blast Radius

The transition to Risk Aware maturity can be executed in 30–60 days once Measured foundations exist.

This timeline assumes an organization with an existing exposure register, executive sponsorship, and the authority to convene cross-functional stakeholders.

Organizations lacking these prerequisites should expect discovery delays rather than execution delays. The work does not become harder—but it becomes political.

Step 1: Define Critical Business Functions

Specific:

- Identify 5–10 non-negotiable business capabilities:
    - Revenue generation
    - Authentication and identity
    - Payment processing
    - Customer communication
    - Regulatory reporting

Measurable:

- One documented capability map

Time-bound:

- 5 business days

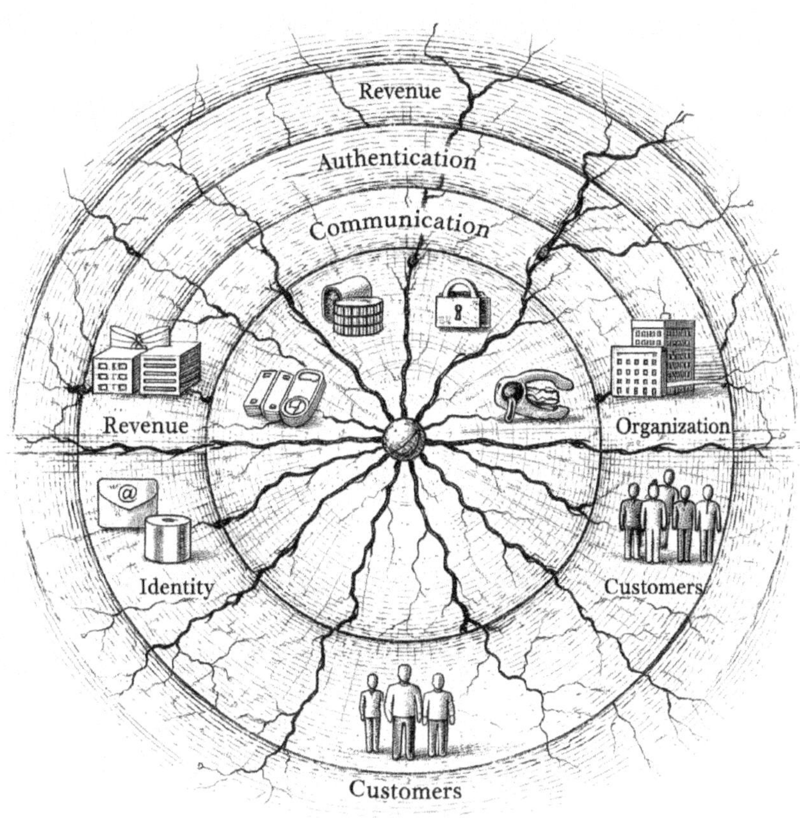

*Blast radius is the scope of harm a single failure
can propagate.*

Step 2: Map Dependencies to Business Functions

Specific:

- For each dependency in the exposure register, identify:
    - Which business functions depend on it
    - Whether that dependency is exclusive or optional

Measurable:

- 100% dependency-to-business mapping coverage

Time-bound:

- 10 business days

Step 3: Classify Blast Radius

Define four blast radius classes:

- **Class 1 – Localized**: Feature degradation only
- **Class 2 – Functional**: One core business function impaired
- **Class 3 – Business**: Revenue or trust interrupted
- **Class 4 – Existential**: Organization-wide failure

These classes represent a minimum viable taxonomy.

Regulated industries may require additional subclasses to capture legal, safety, or ethical consequence, but such extensions refine the model—they do not replace it.

The core distinction remains unchanged: blast radius measures how much of the organization is harmed, not how frequently failure occurs.

Measurable:

- Every dependency assigned exactly one class

Time-bound:

- 5 business days

Step 4: Introduce Recovery Feasibility

Recovery feasibility answers:

Can this damage be undone in time to matter?

Define three recovery tiers:

- **Rapid**: < 1 hour
- **Constrained**: 1–24 hours
- **Trapped**: > 24 hours or contractually blocked

Measurable:

- Each dependency assigned one tier

Time-bound:

- 5 business days

Step 5: Generate the First Risk Ranking

Specific:

- Combine:
    - Exposure tier
    - Blast radius class
    - Recovery feasibility tier
- Produce a ranked Top 10 survivability risk list

Measurable:

- One executive-readable artifact

Time-bound:

- 5 business days

Attribute	Fulfilled
Specific	Business mapping, blast radius classes, recovery tiers
Measurable	Ranked risk list, full dependency coverage
Attainable	No agents, no code changes
Results Oriented	Business-aligned risk prioritization
Time Bound	30–45 days end-to-end

Risk Awareness Execution Summary

SMART Summary

What Changes After Risk Aware

After this step, organizations can say:

> "We know which failures we can tolerate and which ones would destroy us."

That sentence alone defines Risk Aware maturity.

Risk discussions move:

- From engineering rooms to executive rooms
- From opinion to defensible prioritization
- From reaction to intent

The Bridge to Managed

Risk Aware organizations understand danger. Managed organizations redesign to reduce it.

The next chapter moves from:

> "We understand our blast radius."

To:

> "We are changing architecture so it cannot be catastrophic."

That is where survivability becomes design—not hope.

20

From Risk Aware to Managed: Making Risk Architectural

Risk Aware organizations understand what could destroy them. Managed organizations redesign so it cannot.

Risk Aware is knowledge. Managed is action.

This is the most consequential transition in the SPRM maturity model, because it is where performance risk stops being analytical and becomes architectural.

From this point forward, risk is no longer a report. It becomes a design constraint.

This transition typically takes 60–120 days – not because it is technically difficult, but because it forces organizations to confront tradeoffs they have historically avoided.

Risk is not reduced by awareness. It is reduced by design.

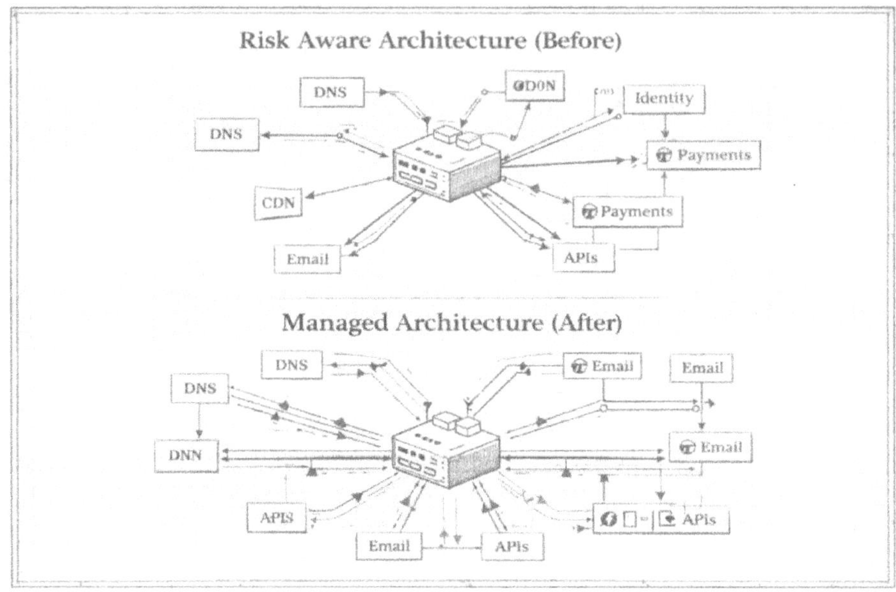

Figure 20.1: Architecture is the control plane of survivability.

What "Risk Aware" Looks Like

A Risk Aware organization has:

- A ranked survivability risk register
- Blast radius classifications
- Recovery feasibility tiers
- Executive visibility into existential risks

They can answer:

- Which dependencies are most dangerous
- Which failures are intolerable
- Where survivability is weakest

But their posture is still observational:

"We know what could hurt us."

Knowledge alone does not change outcomes.

Until architecture changes, exposure remains intact.

What "Managed" Means

Managed means:

> "We are actively engineering survivability into the system."

At this level:

- Survivability becomes an architectural property
- Design decisions require risk justification
- Risk reduction competes for budget
- Exposure reduction becomes measurable work

Risk is no longer observed. It is governed through design.

This is where SPRM stops being advisory.

Architecture as a Risk Control System

At the Managed level, architecture becomes the primary mechanism for reducing exposure.

This means:

- Architecture reviews explicitly reference SPRM findings
- Design documents declare how blast radius is reduced
- Dependency choices require survivability justification

Architecture stops being optimized solely for:

- Performance
- Cost
- Delivery speed

And expands to include:

- Survivability
- Exit feasibility
- Dependency diversity
- Failure containment

This does not slow organizations down. It prevents them from scaling fragility.

The Transition: From Risk Aware to Managed

This transition has six concrete, enforceable steps.

Step 1: Establish a Survivability Design Policy

Specific:

- All new architectures must explicitly declare:
    - Dependency concentration
    - Recovery paths
    - Blast radius reduction intent

Measurable:

- Policy approved, published, and referenced in design reviews

Time-bound:

- 10 business days

Step 2: Integrate SPRM into Architecture Reviews

Specific:

- Add a mandatory Survivability section to:
    - Design documents
    - RFCs
    - ADRs

Measurable:

- 100% of new designs include survivability analysis

Time-bound:

- 15 business days

Step 3: Create a Risk Remediation Backlog

Specific:

- Translate the Top 10 survivability risks into:
    - Architecture changes
    - Dependency diversification efforts
    - Exit path engineering

Measurable:

- Backlog items created, owned, and prioritized

Time-bound:

- 10 business days

Step 4: Define Risk Reduction Metrics

Specific:

- Track progress on:
    - Elimination of single points of failure
    - Increased provider diversity
    - Reduction in recovery constraints

Measurable:

- Baseline established with quarterly improvement targets

Time-bound:

- 10 business days

Step 5: Fund Survivability Improvements

Specific:

- Allocate explicit budget for:
    - Redundancy
    - Dependency exit paths
    - Architectural decoupling

Measurable:

- Budget line items approved

Time-bound:

- Next planning cycle

Step 6: Perform Risk Closure Reviews

Specific:

- Reassess remediated risks quarterly
- Validate that blast radius has actually been reduced

Measurable:

- Risk register updated with closure status and evidence

Time-bound:

- Quarterly cadence

SMART Summary

What Changes After Managed

After Managed maturity:

- Survivability is engineered, not assumed

Attribute	Fulfilled
Specific	Policies, backlog, metrics, funding
Measurable	Risk reduction and closure evidence
Attainable	Uses existing engineering workflows
Results Oriented	Architecture actively reduces exposure
Time Bound	60–120 days initial implementation

- Risk reduction work is visible and funded
- Architectural decisions become defensible under scrutiny

The organization can now state:

"We are actively reducing our risk, not just documenting it."

Why This Is the Hardest Transition

This step requires:

- Political capital
- Budget authority
- Willingness to challenge past decisions

It is easier to acknowledge risk than to change architecture.

Many organizations stall here.

Those that proceed stop being lucky.

The Bridge to Governed

Managed organizations engineer survivability. Governed organizations make survivability non-negotiable.

The next chapter moves from:

"We design for survivability."

To:

"We are accountable for survivability."

That is where SPRM becomes governance.

21

From Managed to Governed: Executive Ownership of Survivability

Managed organizations engineer survivability. Governed organizations *own* survivability.

Managed is operational discipline. Governed is institutional responsibility.

This chapter describes the final and irreversible transition in the Software Performance Risk Management maturity model: the point at which survivability ceases to be an engineering initiative and becomes a formal governance obligation.

This transition does not change tools. It changes authority.

It does not require better dashboards. It requires accountability.

This transition typically takes 90–180 days because it reassigns responsibility at the level where organizations are legally, financially, and reputationally bound.

Engineering can reduce risk. Only leadership can own it.

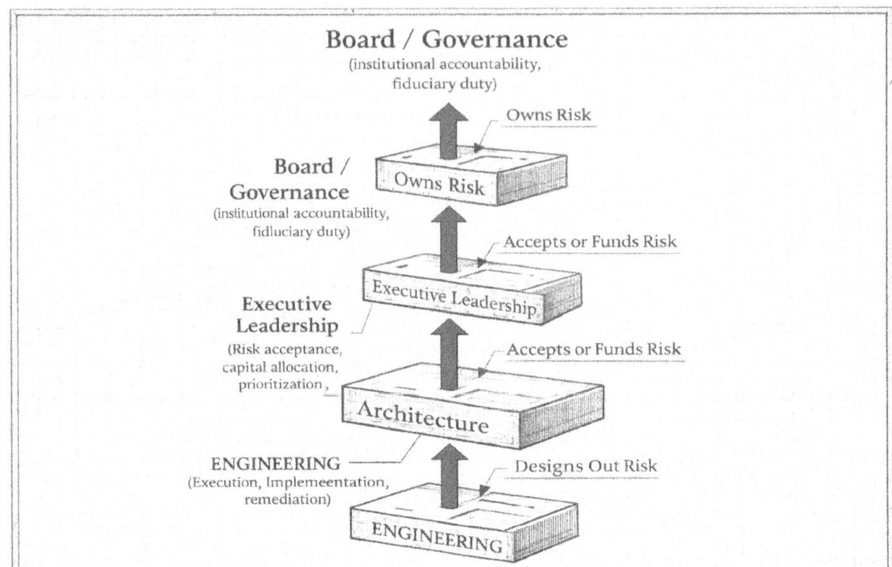

Figure 21.1: Risk without authority is theater. Governance begins where ownership begins.

What "Managed" Really Means

A Managed organization has crossed an important threshold.

It:

- Engineers survivability into system design
- Maintains a ranked survivability risk register
- Integrates SPRM findings into architecture reviews
- Funds remediation work
- Measures reduction in exposure over time

Risk is no longer invisible. Exposure is actively reduced.

But ownership still lives primarily inside engineering and architecture.

The posture is:

"We are fixing risk."

Not yet:

"We are accountable for it."

This distinction matters.

Engineering can reduce exposure, but it cannot legitimately accept business risk on behalf of the enterprise. Engineering does not control capital allocation, regulatory posture, or strategic tradeoffs.

Until those decisions are owned by leadership, survivability remains conditional.

What "Governed" Means

Governed means:

"Survivability is an explicit executive and board-level responsibility."

In a Governed organization:

- Executive ownership of performance risk is formally assigned
- Survivability posture is reviewed at board or risk committee level
- Risk appetite and tolerance are explicitly defined
- High-risk exposure requires conscious, time-bound acceptance
- Survivability informs strategic and capital decisions

Risk is no longer managed by goodwill, heroics, or informal consensus. It is governed by authority.

This is the point at which SPRM stops being a program and becomes policy.

Authority Boundary

Governance in SPRM does not prescribe architecture, tooling, or implementation detail. It constrains acceptable risk.

Engineers remain free to innovate within those constraints. Architects remain free to design. Operators remain free to optimize.

What changes is not creativity, but accountability. Decisions that create existential exposure must be owned by those empowered to accept or reject that exposure on behalf of the organization.

Why Governance Is Necessary

Engineering disciplines can:

- Identify exposure
- Reduce blast radius
- Improve recovery feasibility

They cannot:

- Accept existential exposure
- Trade risk against growth
- Decide which failures are tolerable
- Declare institutional priorities

Only leadership can do that.

Once software performance failures create material financial, regulatory, or reputational harm, survivability becomes inseparable from fiduciary duty. Boards cannot delegate accountability for survivability without accepting liability for its failure.

Without governance, survivability improvements remain optional. With governance, they become mandatory.

Disqualifier

An organization is *not* Governed if any of the following are true:

- High-risk survivability exposure can persist indefinitely without executive sign-off

- Architecture decisions are not constrained by declared risk tolerance
- Risk acceptance is implicit, undocumented, or permanent
- Survivability posture is not visible to executive leadership or the board

Governance is proven by constraint, not by documentation.

The Transition: From Managed to Governed

This transition consists of six enforceable steps.

Step 1: Designate an Executive Risk Owner

Specific:

- Assign a single executive role accountable for software performance survivability:
    - CIO, CTO, or COO

Measurable:

- Formal role charter published
- Survivability accountability explicitly stated

Time-bound:

- 10 business days

This role owns outcomes, not reporting.

Step 2: Establish a Performance Risk Committee

Specific:

- Create a standing committee including:
    - Engineering leadership
    - Security leadership
    - Legal or compliance
    - Finance or enterprise risk

Measurable:

- Committee charter approved
- Regular meeting cadence defined

Time-bound:

- 15 business days

The committee exists to surface tradeoffs, not to dilute accountability.

Step 3: Define Risk Appetite and Risk Tolerance

Specific:

- Establish formal thresholds for:
    - Acceptable blast radius
    - Maximum recovery time
    - Unacceptable dependency concentration

Measurable:

- Policy approved by executive leadership
- Thresholds referenced in design and investment decisions

Time-bound:

- 20 business days

If tolerance is undefined, governance does not exist.

Step 4: Create a Board-Level Survivability Report

Specific:

- Deliver a recurring report that includes:
    - Current survivability posture
    - Changes since prior period
    - Open existential risks
    - Risk remediation progress
    - Explicit risk acceptances

Measurable:

- First report delivered and reviewed

Time-bound:

- 30 business days

If the board cannot see survivability posture, it cannot govern it.

Step 5: Formalize Risk Acceptance

Specific:

- Any unresolved high-risk exposure must:
    - Be explicitly accepted by the executive risk owner
    - Be time-bound with a mandatory expiration date
    - Include documented rationale and compensating controls
    - Be re-approved or remediated upon expiration

Measurable:

- Risk acceptance register maintained and audited

Time-bound:

- 15 business days

Risk that can be accepted forever is not governed.

Step 6: Integrate SPRM into Strategic Planning

Specific:

- Survivability considerations influence:
    - Vendor selection
    - Mergers and acquisitions
    - Market expansion
    - Capital investment

Measurable:

- SPRM referenced in planning and approval artifacts

Time-bound:

- Next planning cycle

Governance exists only when risk shapes strategy.

SMART Summary

Attribute	Fulfilled
Specific	Roles, committees, policies, reports
Measurable	Risk posture, acceptance logs, remediation status
Attainable	Built on existing governance structures
Results Oriented	Survivability owned by leadership
Time Bound	90–180 days

What Changes After Governed

After this transition:

- Survivability becomes a standing governance concern
- Risk tradeoffs are explicit and documented
- Architecture aligns with declared tolerance
- Surprise becomes rare

The organization can truthfully state:

"Survivability is our responsibility, not our hope."

Limitation

Governance does not eliminate failure. It does not guarantee uptime. It does not prevent loss.

It ensures that loss is:

- Anticipated
- Bounded
- Owned

This is the highest standard any institution can meet.

Why This Is the Highest Maturity Level

Before governance:

- Risk is managed locally
- Improvements are discretionary
- Accountability is diffuse

After governance:

- Risk is owned institutionally
- Survivability is non-negotiable
- Authority matches consequence

This is the point at which SPRM becomes irreversible.

The Storm and the Safe Harbor

At this level, the organization has built its harbor:

- Clear authority
- Clear ownership
- Clear accountability
- Clear survivability boundaries

Storms will still come. Failures will still occur.

But they will not be accidental.

That is the final promise of Software Performance Risk Management.

Part IV

Failure Under Load: Case Studies in Ungoverned Survivability

This part introduces no new theory.

Parts I–III established Software Performance Risk Management as a discipline grounded in exposure, blast radius, recovery feasibility, and governance authority. Part IV exists for a single purpose:

To demonstrate how the absence of that discipline manifests under real-world stress.

The chapters that follow are not cautionary tales. They are not postmortems. They are not critiques of individual engineers or organizations.

They are pattern confirmations.

Each case examines a public, high-impact failure through a narrow and deliberate lens:

- What exposure existed before load arrived
- What dependencies were structurally concentrated
- What survivability decisions were never governed
- What authority was absent when tradeoffs mattered

No case relies on privileged information. No case depends on hindsight. No case assumes malicious intent or incompetence.

In every instance, the conditions that produced failure were:

- Known
- Recurrent
- Architecturally visible

Failure occurred not because systems were complex, but because survivability was never owned.

These chapters should be read diagnostically, not narratively. If a failure appears "obvious" in retrospect, that is the point. If a failure appears "unfair" given circumstances, that is the risk being activated, not created.

Part IV does not argue that these failures were preventable in every detail. It argues something more precise:

They were governable.

That distinction is the boundary between bad luck and unmanaged risk.

The reader is invited to test each case against the SPRM framework introduced earlier.

If a failure cannot be explained by ungoverned exposure, concentrated blast radius, or absent authority, then SPRM does not apply.

If it can, then survivability was not a mystery. It was a choice left unmade.

22

Crisis Load: When Systems Are Forced Outside Their Design Envelope

Most software systems are designed for chaotic access patterns.

Unemployment systems, emergency aid portals, public benefits platforms, and regulatory filing systems are built to handle dozens—or at most hundreds—of users arriving within narrow windows. Their access patterns are irregular, bursty, and unpredictable, but historically bounded.

They are not built for synchronization.

During COVID-19, those systems were thrust—simultaneously, globally, and without warning—into a radically different operating regime. What had once been chaotic access became synchronized mass arrival. Systems designed for intermittent demand were suddenly subjected to the digital equivalent of a worldwide flash sale.

Thousands. Tens of thousands. Sometimes millions of users arriving within minutes.

181

This was not a traffic spike. It was a design envelope violation.

Software Performance Risk Management names this condition **crisis load**.

Crisis Load Is Not Scale

Crisis load is not simply "more traffic."

Traditional scale assumes:

- Gradual ramps
- Predictable growth
- Load testing within known parameters

Crisis load violates all three.

It is characterized by:

- Sudden synchronization of demand
- Near-total user intent alignment
- Zero tolerance for delay or retry
- Human consequence attached to every failed interaction

These are not engineering failures. They are survivability failures.

The lesson of COVID is not that systems should have been designed for pandemics. It is that survivability outside assumed operating conditions was not governed.

A Global Pattern, Not a Local One

Systems failed across the civilized world.

The United States. Canada. Europe. Israel. India. Australia. Japan.

Everywhere software mediated access to employment, healthcare, aid, or identity, systems slowed, degraded, or failed under unprecedented stress.

Outcomes were uneven. Some systems degraded gracefully. Others collapsed temporarily. Every system experienced strain.

This chapter does not attempt comparative national analysis. It isolates a shared failure mode observable across jurisdictions regardless of outcome.

The pattern was global. The severity varied.

And in that variation, one case stood out.

Florida: Severity Without Uniqueness of Cause

Florida's unemployment system failure was not unique in kind. It was unique in severity.

The system failed more completely, more visibly, and for longer durations than most of its peers. Citizens were locked out entirely. Applications could not be submitted. Status could not be checked. Human fallback channels were overwhelmed or unavailable.

What makes this case instructive for SPRM is not blame. It is contrast.

Florida's system had been redesigned and rebuilt only a few years prior to COVID. It was not a legacy mainframe. It was not an unmaintained relic.

It was modern.

That matters.

The redesign reflected modernization priorities common across jurisdictions:

- Digital-first access
- Fraud controls
- Cost containment
- Operational efficiency

What it did not reflect was governance of survivability outside assumed demand bounds.

This was not a failure of engineering competence. It was a failure of accountability.

Operating Outside the Assumed World

Like many public systems, Florida's unemployment platform was designed for:

- Regional events
- Seasonal fluctuations
- Localized economic stress

It was not designed for:

- Nationwide simultaneous job loss
- Mandatory digital-only access
- Human desperation attached to every request

When COVID arrived, the system did not merely slow. It exhausted critical resources:

- Session state
- Backend processing capacity
- Queue depth
- Human remediation paths

Recovery was not quick because recovery had not been designed.

This is the core SPRM insight:

> Failure does not reveal risk posture. It reveals who was accountable for survivability outside assumptions.

What SPRM Asks That Postmortems Do Not

Traditional postmortems ask:

> Why did the system fail?

SPRM asks:

> Who owned survivability when the system was forced beyond its design envelope?

SPRM does not require prediction of pandemics. It requires explicit ownership of the question:

> What happens if demand synchronizes?

In Florida's case—as in many others—no authority existed to ask, answer, or constrain that question.

Exposure was implicit. Ownership was diffuse. Survivability was assumed.

Crisis Load as a Governance Failure

Crisis load is not an edge case. It is a recurring pattern.

Draft rumors cause Selective Service systems to crash. Emergency aid portals fail after natural disasters. Healthcare enrollment systems collapse during deadline surges. Commercial platforms fail during flash sales and major announcements.

In commercial systems, these patterns are well understood. In public systems, they are rarely governed.

This is not an engineering gap. It is a governance gap.

The SPRM Doctrine of Crisis Load

SPRM introduces a non-negotiable principle:

> **Axiom**
>
> If a system mediates access to livelihood, safety, or legal standing, then survivability under synchronized demand is a governance responsibility, not an engineering afterthought.

Florida's failure was not that it lacked technology. It was that no one was accountable for survivability beyond assumed demand.

That is what made the outcome catastrophic.

Why This Case Matters

This case is not about Florida. It is about every organization that assumes:

- Demand will remain statistically independent
- Users will arrive gradually
- Retry is acceptable
- Delay is tolerable

Crisis load breaks all four assumptions at once.

SPRM exists to ensure that when assumptions fail, responsibility does not.

Closing

COVID did not create fragile systems. It exposed them.

Florida's unemployment system did not fail because it was old. It failed because survivability outside its design envelope was unowned.

That is the lesson SPRM preserves.

Not to assign blame. But to ensure that the next time the world synchronizes demand, systems fail gracefully—or not at all—because someone was accountable for the question in advance.

That is what survivability governance means.

The only thing more expensive than education is ignorance."

Benjamin Franklin

23

The Cost of Not Knowing

Most catastrophic performance failures are not caused by unpredictable demand.

They are caused by predictable demand applied to systems that were never designed to survive it.

The cost of not knowing is not technical embarrassment. It is wasted capital, lost trust, regulatory exposure, and reputational damage that cannot be insured away.

This chapter examines a recurring pattern in modern software failures: organizations knowingly amplifying demand without knowing whether their systems are survivable.

Predictable Demand Is Not a Stress Test

Modern systems routinely fail under conditions that were fully foreseeable.

In many cases, the date, time, audience size, and behavioral intent of users are known months or years in advance. These events are not chaotic in origin. They are deterministic.

Examples include:

- National benefit enrollment windows
- Emergency relief applications
- Selective Service announcements
- Major marketing events
- Global advertising campaigns

What changes is not the uncertainty of demand, but the willingness of systems to absorb it.

Failure under these conditions is not an accident. It is a design outcome.

The Global Flash-Sale Antipattern

Many public-facing systems are designed for diffuse, asynchronous access:

- Dozens or hundreds of users per hour
- Broad arrival windows
- Gradual session churn

During extraordinary events, those same systems are subjected to:

- Thousands to tens of thousands of arrivals
- Narrow, synchronized time windows
- Immediate transactional intent

This is the functional equivalent of a global e-commerce flash sale.

Yet unlike mature commerce platforms, many of these systems:

- Pre-allocate finite server resources to all visitors
- Fail to distinguish revenue or mission-critical paths
- Treat non-committed users as equals in resource priority

This violates a fundamental survivability principle:

Never pre-allocate scarce resources to a population where the majority will not complete a critical transaction.

When this principle is ignored, both performance degradation and outage risk increase simultaneously.

The Annual Proof: Super Bowl Demand Failures

Since the rise of the modern smartphone era, one pattern has repeated with striking regularity.

Each year, at least one newly advertised website associated with the Super Bowl fails under load.

This is not folklore. It is a structural inevitability.

The Super Bowl produces:

• A fixed date known years in advance

• A synchronized global audience

• An explicit call to action

• Immediate behavioral response

There is no ambiguity about whether demand will surge. Only about whether the system was designed to survive success.

The recurring failure is not caused by extraordinary demand. It is caused by extraordinary optimism.

Organizations invest tens of millions of dollars to generate attention, while investing comparatively little to validate survivability. The result is a failure not of engineering skill, but of governance.

The COVID Shock Was Not Unique — The Severity Was

The COVID era exposed this same pattern globally.

Unemployment systems, healthcare portals, relief applications, and emergency services were subjected to demand far beyond their original design assumptions. Failures occurred across the United States, Canada, Europe, Israel, India, Australia, Japan, and beyond.

Every system experienced stress. Many experienced slowdown. Some experienced partial outage.

What distinguished the most severe failures was not the presence of demand, but the absence of survivability engineering.

In several cases, systems that had been recently redesigned still failed catastrophically. The guidance available to government agencies—both internal and through experienced contractors—had already addressed many of these risks.

The failures were not due to ignorance of best practices. They were due to design decisions that constrained scalability, concentrated dependencies, and assumed orderly access in a disorderly world.

This is not hindsight bias. It is an exposure analysis.

Healthcare.gov Is a Warning, Not a Punchline

Some systems are difficult to reference because of the emotions they evoke.

Healthcare.gov is one such system.

Its early failures were not caused by unprecedented demand. They were caused by ignoring known realities:

- Synchronized enrollment windows
- High user anxiety
- Mandatory participation
- Finite backend resources

The lesson was never that government systems cannot scale. The lesson was that survivability must be designed before demand is amplified.

That lesson has been relearned repeatedly.

The Question That Was Never Asked

Across these failures, one question was consistently absent:

"What happens if this works?"

Not:

- How fast is it under normal load?
- How many users can it handle in testing?

But:

- What is the blast radius if it saturates?
- Which dependencies become chokepoints?
- Who has authority to degrade, shed, or redirect load?

Without answers to these questions, success becomes dangerous.

The Cost of Not Knowing

The cost is not limited to downtime.

It includes:

- Lost revenue from abandoned transactions
- Reputational damage amplified by social media
- Emergency remediation costs
- Regulatory and political fallout
- Long-term erosion of trust

These costs compound.

They are paid not because systems were stressed, but because exposure was invisible.

SPRM's Claim

Software Performance Risk Management does not promise that systems will never slow.

It promises something more important:

> That organizations will know, in advance, whether success is survivable.

Failure under unknown conditions is misfortune. Failure under known conditions is negligence.

The difference is governance.

This chapter exists to make that distinction impossible to ignore.

24

The Non-Delegability Principle

Every large failure eventually produces the same explanation:

"That wasn't my responsibility."

This statement is almost always true.

And that is precisely the problem.

Survivability failures do not occur because people refuse responsibility. They occur because responsibility was never explicitly assigned.

This chapter defines the Non-Delegability Principle: the idea that certain classes of risk—specifically survivability and performance risk—cannot be safely delegated, outsourced, or implied. They must be explicitly owned, governed, and enforced.

Not because people are malicious. Not because teams are negligent. But because human systems do exactly what they are designed to do when responsibility is ambiguous.

The Non-Delegability Principle

> **Axiom**
>
> **Non-delegable risk** is risk whose consequences cannot be transferred, even if its execution is.

Execution can be outsourced. Implementation can be delegated. Operations can be contracted.

Consequence cannot.

When a system fails under load, the business absorbs the damage:

- Revenue loss
- Reputational harm
- Regulatory exposure
- Public trust erosion

No contract absorbs that impact. No vendor indemnifies credibility. No platform refunds lost legitimacy.

Survivability risk is therefore non-delegable by definition.

The Non-Assignment Problem

Most survivability failures are not caused by bad actors, incompetence, or organizational dysfunction.

They are caused by non-assignment.

This is not a moral failing. It is human nature operating under ambiguity.

Humans act on what they are explicitly accountable for. They defer what is adjacent, shared, or implied. They ignore what is undefined.

In complex organizations, this behavior is not pathological. It is adaptive.

When responsibility is unclear, individuals protect their bounded role. When authority is diffuse, decisions are deferred. When consequence is abstract, action is postponed.

Survivability risk is uniquely vulnerable to this pattern.

It does not live cleanly in any single domain:

- It is not purely engineering

- It is not purely security

- It is not purely operations

- It is not purely vendor management

Because survivability spans domains, it is often assumed to be owned by *someone else*.

Not out of avoidance. Out of non-assignment.

The predictable result is this:

> Everyone believes the risk is being handled. No one is responsible for ensuring that it is.

The outcome is not negligence. It is silence.

And silence is how survivability risk accumulates until failure activates it.

Why "Not My Job" Is Usually Accurate

After a failure, organizations conduct postmortems. They identify contributing factors. They ask who should have caught the issue.

The answer is often uncomfortable:

> No one was explicitly responsible for that risk.

This is not evasion. It is an accurate description of the organizational design.

Performance risk often exists:

- Between teams
- Across vendors
- Outside formal charters
- Below executive reporting thresholds

In these gaps, risk is not contested. It is ignored.

Not because it is unimportant. But because it is unowned.

SPRM exists to eliminate these gaps.

Delegation Versus Ownership

Organizations regularly confuse delegation with ownership.

They believe:

- A cloud provider owns scalability
- A CDN owns availability
- A SaaS vendor owns performance

What they have delegated is execution.

What they have not transferred is consequence.

When a dependency fails under load, the business still absorbs:

- Customer outrage
- Lost revenue
- Regulatory scrutiny
- Brand damage

Delegation without retained ownership produces a dangerous illusion: the belief that survivability risk has been "handled."

It has not.

It has merely been obscured.

Non-Delegability as a Governance Constraint

The Non-Delegability Principle forces a structural conclusion:

> If a risk cannot be delegated, it must be governed.

Governance means:

- Explicit ownership
- Defined authority
- Enforced constraints
- Accepted accountability

Without governance:

- Risk becomes implicit
- Decisions become fragmented
- Failures become "surprising"

With governance:

- Risk becomes visible
- Tradeoffs become explicit
- Failures become explainable

This is the transition SPRM exists to enable.

Why This Principle Is Foundational

Every subsequent SPRM concept depends on this one.

Without non-delegability:

- Blast radius is ignored
- Exposure is tolerated by default
- Recovery feasibility is assumed

With non-delegability:

- Survivability becomes a first-class concern
- Authority must be assigned
- Silence becomes unacceptable

This chapter establishes the psychological and organizational reality upon which the rest of the discipline is built.

Survivability risk does not fail because people do not care. It fails because no one was told it was their job.

Closing

The Non-Delegability Principle is not accusatory. It is clarifying.

It explains why intelligent, well-intentioned organizations fail in predictable ways. And it makes the solution unavoidable:

If survivability matters, someone must own it.

SPRM begins at the moment an organization stops assuming that responsibility will emerge—and assigns it instead.

25

Responsibility That Cannot Be Assigned

Every major system failure eventually arrives at the same uncomfortable truth:

No one was responsible.

Not because people were malicious. Not because they were incompetent. But because responsibility was never explicitly assigned.

This is not a technical flaw. It is a human one.

Responsibility That Is Not Assigned Is Not Executed

Human systems obey a simple rule:

If responsibility is not articulated, it is ignored.

This is not apathy. It is survival.

People optimize for:

- Their job description
- Their incentives
- Their evaluation criteria

Anything outside those boundaries becomes:

- Someone else's problem
- A future concern
- An implicit assumption

Software survivability fails most often not because no one cared—but because everyone assumed someone else did.

The "Not My Job" Failure Mode

In post-incident reviews, the same phrases appear repeatedly:

> "That was infrastructure."
> "That was the vendor."
> "That was operations."
> "That was security."
> "That was the business decision."

Each statement is individually defensible.

Collectively, they describe a vacuum.

When responsibility fragments, survivability evaporates.

Why Performance Risk Cannot Be Delegated

Some risks are delegable:

- Implementation
- Optimization
- Monitoring

Survivability risk is not.

It spans:

- Architecture
- Dependency selection
- Budget allocation
- Launch timing
- Business ambition

No single engineering team owns all of these levers.

Therefore:

> Survivability risk is non-delegable.

It must be owned where authority exists.

SPRM as the Assignment of Responsibility

Software Performance Risk Management exists to make one thing explicit:

> Someone is accountable for survivability.

Not advisory. Not suggestive. Accountable.

SPRM does not ask who caused the failure. It asks who accepted the exposure.

That question changes behavior.

The Bridge Forward

Once responsibility is explicit, something new becomes possible:

> The ability to say "No."

The next chapter defines when – and why – SPRM is obligated to exercise that authority.

26

When SPRM Can Say "No"

Most governance frameworks focus on approval.

SPRM must also define rejection.

If survivability risk can be identified, measured, and ranked—but never constrained—then governance is theater.

This chapter defines the conditions under which Software Performance Risk Management must say "No."

"No" Is Not Anti-Innovation

SPRM does not exist to slow delivery. It exists to prevent preventable collapse.

A procedural "No" does not mean:

- The idea is bad
- The team is incompetent
- The business is wrong

It means:

The survivability conditions are unacceptable.

When SPRM Is Obligated to Intervene

SPRM must intervene when all three conditions are present:

- The failure mode is predictable

- The blast radius is existential

- The mitigation requires architectural change

This is not hindsight governance. It is foresight enforcement.

The Pre-Allocation Fallacy

One survivability failure pattern appears repeatedly across industries, decades, and platforms.

The pre-allocation fallacy occurs when finite, stateful system resources are allocated to users who have not yet demonstrated intent, under conditions of unbounded arrival rate.

Commonly pre-allocated resources include:

- Shopping carts

- User sessions

- Server-side memory

- Compute reservations

Under normal conditions, this inefficiency is invisible. Under surge conditions, it is catastrophic.

This is not a performance issue. It is a survivability violation.

Case Study: Nike and the Air Yeezy 2 Launch

In 2012, Nike launched the Air Yeezy 2 collaboration with Kanye West.

The inventory was approximately 500 units. The demand was effectively unbounded.

Sneaker collectors, fans, resellers, and cultural observers arrived within a narrow time window. Nike's e-commerce platform allocated stateful sessions and shopping carts to every arriving visitor.

The result was predictable:

- Finite server resources were consumed by non-converting users
- High-intent buyers competed with curiosity traffic
- The site locked for hours
- Load dissipated only through session expiration waves

This was not a scaling failure. It was a design failure.

The system behaved exactly as designed.

That is the indictment.

Why This Was Governable Before Launch

All relevant facts were known in advance:

- Inventory was scarce
- Demand was extreme
- Arrival rate was unbounded
- Conversion probability was tiny

Under SPRM, this architecture would have triggered a procedural "No."

Not because the launch was risky—but because it violated a survivability constraint:

> Finite resources were pre-allocated to a population where the majority would never complete the mission.

What SPRM Requires Instead

SPRM does not prohibit launches. It requires survivability-aware design:

- Intent gating before state allocation
- Protection of critical transaction paths
- Explicit rejection of curiosity traffic under surge

When these conditions are unmet, SPRM must either:

- Require architectural redesign
- Or demand explicit executive risk acceptance

Silence is not an option.

The Ethical Dimension

When systems mediate access to:

- Employment
- Healthcare
- Financial relief
- Civic participation

Failure is not merely technical.

Governance without ethics is administration. Ethics without governance is aspiration.

SPRM occupies the space where responsibility becomes moral.

Closing Doctrine

SPRM does not say "No" often.

But when it does, it speaks for:

- Users who cannot retry endlessly
- Citizens who depend on access
- Organizations whose legitimacy depends on reliability

A system that cannot say "No" before launch will eventually fail after it.

SPRM exists to ensure the "No" comes first.

27

The Lantern

There is an old story about Diogenes.

He walked the streets of Athens in daylight carrying a lantern. When asked what he was doing, he replied that he was looking for an honest man.

The point was not that honesty did not exist. The point was that it could not be found without light.

This chapter exists because governance without illumination becomes ritual, and illumination without authority becomes noise.

Software Performance Risk Management does not exist to judge intent. It exists to make consequence visible.

Illumination, Not Absolution

The lantern was not a symbol of virtue. It was a tool of exposure.

Diogenes did not offer forgiveness. He did not offer solutions. He did not offer comfort.

He revealed what already was.

SPRM plays the same role.

It does not accuse individuals. It does not assign moral worth. It does not promise perfection.

It exposes where systems silently decide who gets access, who waits, and who is excluded when failure occurs.

Once that exposure is visible, responsibility becomes unavoidable.

Governance Begins Where Excuses End

Before illumination, every failure has an excuse:

> "No one could have predicted this."
> "The demand was unprecedented."
> "That system was never designed for this use."
> "That dependency belongs to another team."

The lantern does not argue with these statements. It renders them irrelevant.

Once exposure is visible, the only remaining question is:

> "Who is authorized to accept this consequence?"

If no one is authorized, governance does not exist.

Ethics as an Emergent Property of Authority

This is the point at which governance and ethics collide.

Not because ethics were invited. Because they cannot be excluded.

When software mediates access to income, healthcare, identity, education, or safety, survivability decisions become ethical outcomes regardless of intent.

This is not philosophy. It is causality.

Authority without accountability produces harm. Visibility without ownership produces theater.

SPRM exists to prevent both.

The Lantern Does Not Judge

The lantern does not care why a decision was made. It does not care who was under pressure. It does not care what incentives existed at the time.

It reveals only this:

- Where survivability depends on fragile assumptions
- Where failure concentrates human impact
- Where responsibility has not been assigned

What happens next is no longer technical.

It is a governance decision.

The Irreversible Moment

Once the lantern is raised, ignorance is no longer available as a defense.

The organization can still choose risk. It can still accept exposure. It can still prioritize speed, cost, or convenience.

But it must do so consciously, explicitly, and with authority.

Once the lantern is raised, the only remaining choice is whether governance follows it — or whether exposure is left to become harm.

28

Software Performance Risk Management as a Continuous Discipline

Every discipline that matters is continuous.

Finance is continuous. Security is continuous. Compliance is continuous.

Survivability must be continuous.

If Software Performance Risk Management is treated as a project, it will fail. If it is treated as a report, it will be ignored. If it is treated as a tool, it will be replaced.

SPRM only succeeds when it becomes a discipline.

This chapter defines how Software Performance Risk Management remains alive, authoritative, and accountable over time—and why it cannot exist without ownership, renewal, and governance.

Why Disciplines Never Finish

Projects end. Disciplines persist.

A project:

- Has a start date
- Has an end date
- Produces an artifact

A discipline:

- Evolves
- Reassesses
- Governs continuously

Risk does not complete. Architecture does not complete. Dependencies do not stabilize.

Survivability, therefore, cannot complete.

Any organization that believes it has *finished* managing performance risk has already begun to accumulate it again.

Governance, Ethics, and the Ancient Problem of Power

Long before software, societies struggled with a familiar problem: how to wield power without causing harm.

Aristotle distinguished between *techne* — the skill to build – and *phronesis* — the practical wisdom to govern consequences.

A system can be expertly constructed and still be unfit to serve its purpose.

Governance was never about morality in the abstract. It was about responsibility in the concrete.

In Greek civic thought, authority without responsibility was not neutral: It was dangerous. Power unaccompanied by accountability produced harm not through malice, but through neglect.

Modern software systems now exercise comparable power:

- They gate access to money
- They gate access to healthcare
- They gate access to employment, aid, identity, and participation

When such systems fail under load, they do not merely degrade service. They impose asymmetric harm.

This is not a moral argument. It is a governance argument.

Any system that claims to support a mission, and becomes a barrier to that mission under stres, —has violated its responsibility to the community it serves.

The Renewal Cycle of SPRM

SPRM operates in cycles, not milestones:

- Discovery
- Measurement
- Prioritization
- Remediation
- Verification
- Renewal

Each cycle:

- Hardens institutional memory
- Reduces survivability exposure
- Clarifies accountability

Governance does not mean preventing change. It means adapting without forgetting.

Institutionalizing Responsibility

For SPRM to remain continuous, it must be embedded in institutions, not individuals.

It must exist in:

- Organizational charters
- Governance policies
- Architecture standards
- Board and risk committee reporting

It must not live inside:

- A single team
- A single tool
- A single champion

Disciplines survive personnel changes. Projects do not.

Cadence Is Authority

Survivability is built through rhythm.

Recommended governance cadence:

- Monthly: Risk register review
- Quarterly: Executive survivability briefing
- Semi-annual: Architecture survivability audit
- Annual: Policy and maturity reassessment

Cadence transforms risk from emergency into governance.

Where cadence lapses, authority erodes.

Model Drift and the Danger of Forgotten Risk

Every model drifts.

Dependencies change. Providers consolidate. Business priorities shift.

SPRM therefore requires:

- Periodic re-baselining
- Model recalibration
- Risk reprioritization

Risk intelligence without renewal becomes superstition.

SPRM as Organizational Memory

Most outages repeat known patterns.

Organizations do not fail because they lack data. They fail because they forget.

SPRM becomes:

- The record of survivability decisions
- The archive of accepted risks
- The justification for architectural constraints

It explains not only *what* systems are built, but *why* they are shaped the way they are.

The Non-Delegability Boundary

There is a boundary beyond which responsibility cannot be delegated.

Engineering can reduce risk. Vendors can mitigate components of risk. Tools can surface signals.

But responsibility for survivability cannot be outsourced.

If a system can deny access to money, healthcare, or legal status under load, and no named executive is accountable for its survivability, then failure is not an accident. It is a governance defect.

Authority without responsibility is not neutrality. It is abdication.

What It Means to Wear the Mantle

SPRM does not claim moral perfection. It claims responsibility.

It accepts that:

- Systems will fail

- Tradeoffs will exist

- Risk can never be eliminated

What it refuses is invisibility.

To wear the mantle of governance is to say:

- We understand the exposure

- We accept or reduce it consciously

- We own the consequences

This is not heroism. It is adulthood.

The Harbor Is Never Finished

A harbor is not built once.

- Docks are repaired

- Walls are reinforced

- Charts are updated

SPRM is harbor maintenance for modern organizations.

Storms are inevitable. Loss is optional.

The Closing Doctrine

Software Performance Risk Management is not a product. It is not a framework. It is not a methodology.

It is a discipline.

A discipline that ensures performance risk is never again invisible, unowned, or unmanaged.

And because software now mediates human lives, this discipline is inseparable from responsibility.

That is the final promise of Software Performance Risk Management.

Minimal Viable SPRM Implementations (MV-SPRM)

Non-normative appendix notice. The implementations in this chapter are illustrative examples demonstrating that Software Performance Risk Management (SPRM) can be instantiated across a range of organizational toolchains and maturity levels. These examples are not requirements, do not prescribe specific vendors or products, and may be adapted without changing the discipline definition presented in the Manifesto introduction.

Purpose and Scope

This chapter provides three Minimal Viable SPRM (MV-SPRM) patterns: (1) predominantly commercial off-the-shelf (COTS), (2) a blend of COTS and open source, (3) COTS + open source + custom components. Each pattern is designed to establish *risk visibility*, *ownership*, and *decision leverage* without requiring a full platform rewrite or organizational reorg.

When to Use MV-SPRM

- You suspect performance risk exists but cannot localize it with evidence.
- You are repeatedly surprised by latency regressions or reliability events tied to performance.
- You have monitoring and testing, but no governance for performance risk ownership and escalation.

What MV-SPRM Must Produce

MV-SPRM is considered present when it reliably produces the following artifacts:

1. A **Performance Risk Register** that is reviewed and maintained.

2. A **Risk Ownership Model** (named humans/teams) for each material surface.

3. **Thresholds and Escalation Criteria** expressed in business-relevant terms.

4. **Evidence Trails** (what was observed, when, and why decisions were made).

Definitions (Local to This Chapter)

Performance Risk Surface Any execution path, dependency, platform, or operational condition whose degradation can create material business impact.

Material Meaningfully consequential to revenue, mission, safety, compliance, customer trust, or operational continuity.

Evidence Observations derived from real system behavior, telemetry, logs, user experience, test results, or controlled experiments sufficient to support a decision.

MV-SPRM Pattern 1: Predominantly COTS

Objective

Establish a risk-governed performance baseline using tools the organization already owns.

Minimum Inputs

- System inventory and critical user journeys (top flows)
- Existing logs/metrics/traces and incident history
- A small number of business outcomes that define material impact

Core Capabilities

1. **Risk Inventory** (journeys, services, dependencies, shared infrastructure)

2. **Baseline Evidence** (distributions, tails, error amplification, time-based patterns)

3. **Thresholds and Escalation** (who is paged, when; what constitutes a risk state)

4. **Ownership** (named accountable owners per risk surface)

Deliverables

- **Performance Risk Register** (initial)
- **Dependency and Boundary Map** (initial)
- **Baseline Report** (30–90 day view if possible)
- **Executive Risk Summary** (one page)

Acceptance Criteria (Exit Conditions)

1. At least one critical journey has a baseline with tail visibility (not averages only).

2. At least one recurring degradation has a named owner and an agreed escalation path.

3. At least one mitigation decision is recorded with evidence.

Known Failure Modes

- Everything becomes a dashboard; nothing becomes a decision.
- Averages are used; tails remain invisible.
- Ownership is implied, not explicit.

MV-SPRM Pattern 2: COTS + Open Source Blend

Objective

Reduce blind spots and improve diagnostic clarity via correlation, normalization, and statistical sufficiency.

Minimum Additions Over Pattern 1

- A normalization layer for event identity and time alignment
- A durable storage pattern for trend analysis
- Basic statistical sufficiency checks (sample size, confidence, noise vs change)

Core Capabilities

1. **Cross-signal Correlation** (logs + metrics + traces + user experience, when available)

2. **Passive Evidence Leverage** (access logs, CDN logs, protocol evidence)

3. **Statistical Confidence** (confidence intervals, precision bands, sufficiency)

4. **Risk Trending** (degradation emergence, seasonal patterns, regression detection)

Deliverables

- **Risk Trends** per critical journey and dependency
- **Attribution Summaries** (what changed, where, and likely contributors)
- **Evidence Packets** attached to risk register items

Acceptance Criteria (Exit Conditions)

1. At least one risk trend is tracked across time with stable identity and timestamps.

2. At least one remediation is prioritized based on evidence of materiality.

3. At least one dependency-boundary issue is detected without a synthetic test.

Known Failure Modes

- Data integration without governance: correlation produces debate, not action.
- Open source becomes a hobby project without owners.

MV-SPRM Pattern 3: COTS + Open Source + Custom Components

Objective

Institutionalize SPRM as a durable capability: risk models, systematic ingestion, and decision automation.

Minimum Additions Over Pattern 2

- A custom risk model aligned to business impact (weights, thresholds, consequences)
- Automated evidence ingestion and classification for common sources
- Governance hooks into release and change processes (reviews, gates, exception handling)

Core Capabilities

1. **Risk Scoring / Banding** tied to material impact definitions

2. **Systematic Ingestion** from heterogeneous sources into normalized events

3. **Decision Automation** (alerts that trigger governance actions, not noise)

4. **Institutional Memory** (historical ledger of risks, mitigations, outcomes)

Deliverables

- **Performance Risk Ledger** (living, versioned, auditable)
- **Risk State Dashboards** designed for decisions (not observability theater)
- **Release/Change Integration** checklists and exception pathways
- **Quarterly Risk Review Pack** suitable for executive review

Acceptance Criteria (Exit Conditions)

1. Risk register items automatically attach evidence packets where feasible.

2. At least one release decision is influenced by quantified risk state.

3. Risk ownership persists across team changes (institutional memory exists).

Known Failure Modes

- Custom components become a platform replacement project.
- Risk scores become a KPI game instead of a decision support system.

Cross-Pattern Governance: The Minimum Controls

Regardless of toolchain composition, MV-SPRM requires the following minimum controls:

Control 1: Ownership

Each performance risk surface must have:

- a named accountable owner,
- an escalation path,
- a review cadence.

Control 2: Thresholds

Thresholds must be:

- stated in business-relevant terms,
- measurable using available evidence,
- paired with actions (not just alerts).

Control 3: Evidence and Decision Traceability

Every material mitigation must record:

- the evidence used,
- the decision made,
- the expected risk reduction,
- the outcome observed (post-change).

Implementation Checklist (One Page)

1. Identify top journeys and their material impact statements.

2. Enumerate dependencies and boundaries (internal + external).

3. Establish baseline distributions (include tails).

4. Define thresholds and escalation actions.

5. Create and maintain the performance risk register.

6. Attach evidence packets and record decisions.

7. Review trends and adjust priorities on a fixed cadence.

Appendix Artifacts (Templates)

Template: Performance Risk Register Entry

Risk ID: ____

System / Journey: _______________________________

Risk Surface: _______________________________

Material Impact Statement: _______________________________

Evidence Sources: _______________________________

Baseline Summary: _______________________________

Threshold / Trigger: _______________________________

Owner / Escalation: _______________________________

Mitigation Options: _______________________________

Decision + Rationale: _______________________________

Outcome / Follow-up Date: _______________________________

Template: Executive Risk Summary (One Page)

- **Top 3 performance risks** (material impact stated)
- **Current risk state** (stable / degrading / critical)
- **Decisions in progress** (what is being done and why)
- **Expected risk reduction** (qualitative or quantified)

Frequently Asked Questions (FAQ)

This appendix exists to address recurring questions that arise whenever Software Performance Risk Management (SPRM) is introduced into mature organizations.

These questions are not misunderstandings. They are boundary probes.

Each reflects an existing discipline attempting to locate SPRM within its familiar frame. The purpose of this appendix is not to defend SPRM, but to establish its jurisdictional boundaries explicitly and permanently.

Isn't this just Site Reliability Engineering (SRE) under a different name?

No.

Site Reliability Engineering governs system behavior during execution. It operates on the assumption that core infrastructural conditions already hold: that names resolve, routes converge, trust validates, and execution can occur.

SPRM governs whether those assumptions are survivable.

SRE answers questions such as:

- Is the system reliable under load?
- Are error budgets being respected?
- Can service-level objectives be met?

SPRM answers a different class of question:

- What happens if execution cannot begin?
- Who controls recovery when trust or routing fails?
- Is failure survivable when instrumentation never runs?

If DNS fails, SRE has no signal. If routing disappears, error budgets are irrelevant. If trust collapses, SLIs do not activate.

SRE operates *inside* survivability boundaries. SPRM governs whether those boundaries exist.

The disciplines are complementary and non-overlapping.

Why should executives care if performance has traditionally been an engineering concern?

Performance failures are no longer technical inconveniences. They are revenue events, reputational events, and frequently legal or regulatory events.

Engineering optimizes execution. Executives govern exposure.

SPRM exists because survivability failures bypass operational response entirely. They halt revenue, sever trust, and remove the organization's ability to act before engineering can intervene.

Boards do not govern dashboards. They govern conditions under which the organization can continue to operate.

If survivability exposure exists without ownership, the failure is already a governance failure—whether or not an incident has occurred.

Isn't this already covered by enterprise risk management (ERM)?

No.

Enterprise Risk Management focuses on probabilistic, financial, and insurable risk categories. It evaluates scenarios, aggregates likelihoods, and allocates capital against loss.

SPRM governs structural survivability conditions that are often:

- non-probabilistic,
- non-insurable,
- non-transferable.

ERM governs outcomes. SPRM governs whether outcomes can be survived.

A system that cannot recover from dependency collapse is not a financial risk problem. It is a structural survivability problem that ERM is not designed to see.

Why doesn't SPRM produce a numeric risk score or maturity index?

Numeric scores imply precision. Survivability failures are discontinuous.

A dependency is not "30% catastrophic." It is either recoverable or it is not.

Governance requires thresholds, not decimals. Executives act on boundaries: acceptable versus unacceptable, survivable versus existential.

SPRM therefore ranks exposure qualitatively by blast radius and recovery feasibility. It does not reduce existential risk to a number that creates false confidence.

A precise score attached to an unbounded risk is still unbounded risk.

How is this different from security or threat modeling?

Security governs malicious intent, exploitability, and unauthorized access. It asks whether a system can be attacked.

SPRM governs dependency, authority, and recovery failure. It asks what happens when a trusted system disappears, fails, or becomes unreachable—without adversarial intent.

A system can be secure and still unsurvivable.

Security protects mechanisms. SPRM governs reliance on them.

These disciplines are orthogonal by design.

Is SPRM anti-cloud, anti-SaaS, or anti-modern architecture?

No.

SPRM is not anti-cloud, anti-SaaS, or anti-modern architecture. It is anti-unexamined dependency concentration.

Cloud platforms, SaaS providers, and managed services increase efficiency while simultaneously increasing blast radius when monoculture or control-plane dependency exists.

SPRM does not require avoidance. It requires exit paths.

Trust without exit is not trust. It is captivity.

Where do observability, APM, and telemetry fit into SPRM?

Observability explains execution. SPRM governs survivability assumptions.

Telemetry activates only after success conditions hold: name resolution, routing, trust, and execution. Many existential failures occur before any signal can be collected.

Observability provides evidence. SPRM provides judgment.

Observability can explain how a system failed. SPRM determines whether failure was survivable.

Closing Note

These questions recur because existing disciplines optimize execution while leaving survivability implicit.

SPRM exists to make survivability explicit, bounded, and owned.

It does not replace engineering, security, or operations. It constrains the conditions under which their work remains viable.

Epilogue: Safe Harbor

Imagine this.

The sun is sinking low over a wide stretch of highway. The desert is still. The sky is burning with oranges and reds, the kind of sunset that feels earned rather than dramatic.

A Sprinter van is pulled off at a quiet rest stop. On its rear door, slightly crooked and unapologetic, is a small Jolly Roger. Not loud. Not theatrical. Just present.

Not a pirate flag of chaos. A pirate flag of responsibility.

A reminder that someone chose to navigate dangerous waters instead of pretending they did not exist. That someone mapped the reefs. That someone built the harbor before the storm arrived.

An IT engineer sits outside the van with a cup of coffee. Boots on the pavement. Elbows resting on their knees. Watching the light fade.

Their phone is in their pocket. It is not vibrating. It is not screaming. It is not demanding to be saved.

A site launch is happening hundreds of miles away.

Customers are flowing through it. Orders are completing. Payments are clearing. Pages are loading. Trust is holding.

No dashboards are flashing red. No war rooms are forming. No one is refreshing social feeds looking for answers.

Nothing dramatic is happening.

And that is the victory.

Because in a world addicted to heroics, silence is mastery. Calm is competence. Boredom is reliability.

This is what safe harbor looks like.

Not the absence of storms, but the presence of preparation. Not perfection, but resilience that feels invisible. Not luck, but ownership exercised long before it was needed.

The Jolly Roger on the van is not rebellion against order. It is allegiance to it.

Not a pirate of chaos. A pirate of order. One who claims responsibility where others prefer plausible deniability. One who accepts that navigation is work, and safety is built, not assumed.

This is what Software Performance Risk Management creates:

A world where launches are not acts of courage. A world where uptime is not a miracle. A world where failure is anticipated, bounded, and governed. A world where fragility is replaced with intention.

The greatest success of SPRM is not that nothing ever breaks. It is that nothing feels fragile.

It is knowing that when something does break, it will not surprise you. It will not cascade. It will not become a story of regret.

It will be handled. Contained. Understood.

This is what maturity looks like.

Not speed. Not power. Not cleverness.

Stewardship.

And stewardship is quiet.

No soundtrack.

No dialogue.

Just the soft scrape of a chair on pavement. The slow sip of coffee. The engine ticking as it cools.

And the western sky putting on its nightly show.

Nothing moves because nothing needs to.

The work was done earlier. The responsibility was accepted long ago.

Now there is only calm.

No soundtrack. No dialogue.

Just the certainty that the job was finished before anyone noticed.

Index

About the Author

James L. Pulley III was born in Spartanburg County, South Carolina. By the sixth grade, he had lived in four cities and learned something that would shape everything that followed: that people can live next to the ocean and never see the beach. Proximity is not awareness. That observation became a career.

He graduated from Furman University in 1991 with a blended degree in computer science and business administration, and began his career at Microsoft, answering questions on the phone for Product Support Services. He became one of Microsoft's early electronic services voices on CompuServe. As an electronic services pioneer, this was the first of many times he would find himself explaining complex systems to people who needed help more than theory.

On April 1, 1996, he was hired as a field sales engineer at Mercury Interactive, the company that defined the performance testing industry. That date, and the discipline it introduced him to, became the origin coordinate of a thirty-year career in software performance engineering. April 1st was chosen for the launch of this book as the thirtieth anniversary of this hire date.

Over three decades, his work has spanned performance testing, capacity planning, application performance management, observability, and site reliability engineering. He has tested systems that governed a trillion dollars in mortgage assets, identified architectural flaws worth millions in monthly revenue, helped secure biometric entry records for every non-citizen entering the United States, and scaled a national digital wallet from ten thousand to one million concurrent users in four weeks. He has built teams, designed governance frameworks, and served as the person organizations called when the system was failing and the stakes were measured in minutes.

But the work he considers most important has always happened in the margins.

In high school, a neighbor's son lost a leg in a tragic accident during his junior year. James was asked to serve as his geometry tutor during recovery. That young man now builds custom prosthetic limbs for others who face the same challenge. The helped became the helper. It was the first time James saw the cycle that would define his professional life, but not the last.

On September 12, 2001, he began answering questions from peers in online technical forums. Not as strategy. Because it was the only useful thing available. Those answers, and the thousands that followed across the next two and a half decades, exceed the length of Homer's Odyssey. They are permanent. They are public. And they are the foundation on which every book, every framework, and every reputation in this body of work was built.

In 2008, he founded Journeyman Publishing — named for the guild tradition in which a journeyman is one who has completed an apprenticeship and earned the right to travel independently, practicing the craft wherever it is needed. He chose the title deliberately. While some of his peers see him as a master, he sees himself as a permanent learner on a journey that does not end.

In 2012, he co-founded PerfBytes, a podcast on software performance engineering inspired by the humor of CarTalk, the accessibility of Alton Brown's Good Eats, and the timing of British comedy. The show has been downloaded more than 177,000 times and continues today. It is proof that technical depth and human warmth are not opposites.

He serves on the board of the Advancing Technology Ventures Lab, a 501(c)(3) founded by Dennis Hayes to mentor early-stage ventures in upstate South Carolina. He has led, mentored, taught, and answered questions in every role he has held. In many cases, this was a task he was never asked to fill.

He often jokes that if he could extract a few cents from every success he has helped others achieve, he would never have to work again. When his father passed in 2006, he received handwritten letters from people around the world whose careers he had quietly shaped. Those letters confirmed what the work had always told him: that usefulness, practiced consistently, is the only authority that survives.

He is the author of *Navigating Performance Engineering*, *Software Performance Risk Management*, *The Fleetwood*, and the *Reputation Engineering* education series. His earlier work includes *Interviewing & Hiring Software Performance Test Professionals*, as well as dozens of articles and whitepapers across forums and conferences.

He lives in Spartanburg County, South Carolina, with his wife Rachel, in a home built around a log cabin from the late 1700s that they are expanding together, by hand. Rachel has been on this land, purchased by her parents, since she was six years old. At the same time James wandered sixteen cities across the country between 1976 and 2018 before arriving at the same three acres. Dorothy had it right in the *Wizard of Oz*, "There's no place like home…"

James has never believed that expertise is the point. The point is the person standing next to you who doesn't have it yet.

www.ingramcontent.com/pod-product-compliance
Lightning Source LLC
Chambersburg PA
CBHW042044030726
47599CB00019B/2355